P9-BVM-851

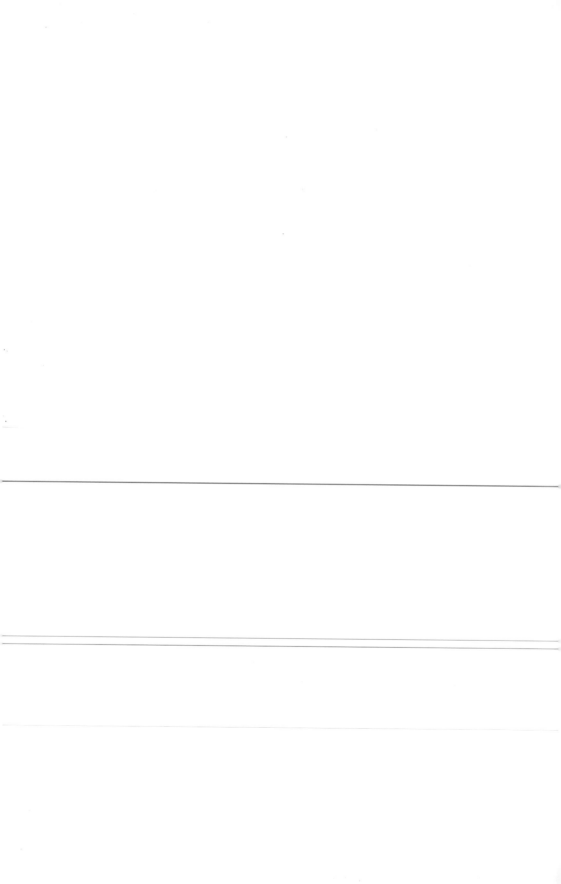

I. Homma, S. Shioda (Eds.)

Breathing, Feeding, and Neuroprotection

I. Homma, S. Shioda (Eds.)

QP
121
AIB743
2006
VAN

Breathing, Feeding, and Neuroprotection

With 47 Figures, including 7 in Color

 Springer

SH

Ikuo Homma, M.D., Ph.D.
Professor and Chairman
Department of Physiology, Showa University School of Medicine
1-5-8 Hatanodai, Shinagawa-ku, Tokyo 142-8555, Japan

Seiji Shioda, M.D., Ph.D.
Professor and Chairman
Department of Anatomy, Showa University School of Medicine
1-5-8 Hatanodai, Shinagawa-ku, Tokyo 142-8555, Japan

ISBN-10 4-431-28774-4 Springer-Verlag Tokyo Berlin Heidelberg New York
ISBN-13 978-4-431-28774-2 Springer-Verlag Tokyo Berlin Heidelberg New York

Library of Congress Control Number: 2006923431

Printed on acid-free paper

© Springer-Verlag Tokyo 2006
Printed in Japan
This work is subject to copyright. All rights are reserved, whether the whole or part of the material is
concerned, specifically the rights of translation, reprinting, reuse of illustrations, recitation, broadcasting,
reproduction on microfilms or in other ways, and storage in data banks.
The use of registered names, trademarks, etc. in this publication does not imply, even in the absence of a
specific statement, that such names are exempt from the relevant protective laws and regulations and
therefore free for general use.
Product liability: The publisher can give no guarantee for information about drug dosage and application
thereof contained in this book. In every individual case the respective user must check its accuracy by
consulting other pharmaceutical literature.

Springer-Verlag is a part of Springer Science+Business Media
springer.com

Typesetting: Camera-ready by the editors and authors
Printing and binding: Asia printing Office Corporation, Japan

Foreword

It is a great pleasure to publish *Breathing, Feeding, and Neuroprotection*. This is a memorable book for Showa University, prepared in coordination with the First International Symposium for Life Sciences, held at Showa University in 2004. This symposium was supported in part by The Special Subsidies (Grants for the promotion of the advancement of education and research in graduate schools) in Subsidies for ordinary expenses of private schools from the Ministry of Education, Culture, Sports, Science and Technology Japan. On behalf of Showa University, it is a privilege to present this book of research findings for the advancement of knowledge of brain functions and morphology.

Akiyoshi Hosoyamada, M.D., Ph.D.
President
Showa University, Tokyo 142-8555, Japan
December 2005

Preface

Brain research is on the march, with several advanced technical developments and new findings uncovered almost daily. Within the brain-research fields, we focus on breathing, neuroprotection, and higher brain functions. One prominent target of respiration research is the central pattern-generation of breathing, in particular respiratory rhythm generation. Where and how respiratory rhythm is generated in the brainstem is of universal interest. This book includes a description of this unresolved problem. We also focus on neuroprotection, with chapters dealing with the functional significance of the blood–brain barrier as an interface of blood and the central nervous system. Several chapters concerning the function of proinflammatory cytokines, pituitary adenylate cyclase-activating polypeptides, and free radicals after brain ischemia are also included. Other chapters have to do with health and disease in relation to brain functions, particularly the hypothalamic and limbic systems. Orexin/hypocretin deficiency is a cause of narcolepsy, and stress induces inhibition of cellular immune responses. We also focus on causes and therapeutic effects of antidepressants from a molecular and cell biology approach.

In addition to animal experiments, this book includes research on the human brain. It is necessary to examine highly developed brain functions in humans. One human-brain research group has been developing the EEG/dipole tracing method. This book includes several chapters dealing with that method, indicating source generators associated with human brain functions.

In addition, this volume contains material from the Showa University International Symposium for Life Sciences, held at Showa University, Tokyo, in August 2004.

Ikuo Homma, Professor of Physiology
Seiji Shioda, Professor of Anatomy
Showa University School of Medicine
January 2006

Contents

Part I Central Regulation of Breathing

Part II Neurogeneration and Neuroprotection

Part III Brain Functions in Health and Disease

Part IV Brain Functions by the Dipole Tracing Method

Color Plates

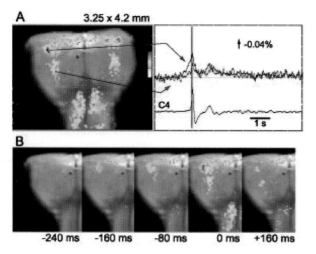

A 3.25 x 4.2 mm

†-0.04%

C4

1 s

B

-240 ms   -160 ms   -80 ms   0 ms   +160 ms

Part I. (by Onimaru et al.) **Fig. 2** (P. 8).

Optical imagings of respiratory neuron activity in the ventral medulla of a wild type mouse in the presence of 10 nM substance P. **A**, *left*: Optical image of respiratory neuron activity near the C4 peak. *Right*: Fluorescence changes at two different points indicated on **A**. **B**: Image is superimposed on the ventral surface of the right half of the medulla. The *numbers* below each image denote the time from the peak of C4 inspiratory activity. Refer to text.

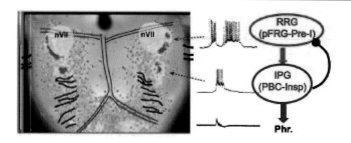

Part I. (by Onimaru et al.) **Fig. 3** (P. 9).

Optical imaging of respiratory neuron activity in the ventral medulla and hypothetical model of respiratory rhythm generation. The optical imaging was obtained from a 0-day-old rat preparation that was stained with a voltage sensitive dye. Refer to text.

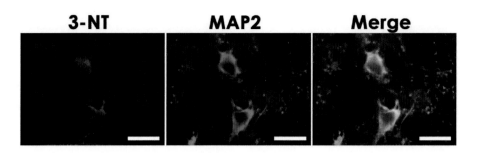

Part II. (by Ohtaki et al.) **Fig. 3** (P. 44).
Oxidative stress increases in neuron after ischemia. 3-nitro-L-tyrosine (3-NT, *red*), which is oxidative metabolite of L-tyrosine by ONOO⁻. The 3-NT positive cell (*red*) is stained observed in MAP2 positive cells (*green*) as neuronal markers. *Bars,* 20 μm.

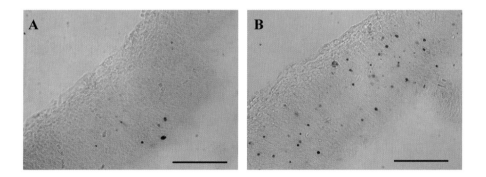

Part II. (by Ogawa et al.) **Fig. 4** (P. 66).
TUNEL staining of fetal mouse brains (embryonic day 12.5) 24 hr after treatment with BrdU. **A**, control. **B**, BrdU. Sections including the neocortex were processed for this staining. Marked TUNEL-positive reactivity was observed in brains treated with BrdU (**B**). *Bars,* 100 μm.

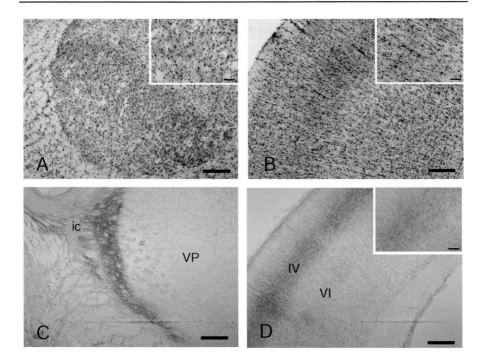

Part III. (by Muneoka et al.) **Fig. 5** (P. 99).
S-100B staining counterstained with cresyl violet in ventroposterior thalamic nucleus (*VP*) and postnatal day (PND) 7 (**A**) and in the somatosensory cortex at PND 15 (**B**).5HTT staining in VP (**C**) and the somatosensory cortex (**D**) at PND 7. *Scale bars,* 200 μm and 50 μm (inlets). *ic*, internal capsule; *IV*, layer IV; *VI*, layer VI.

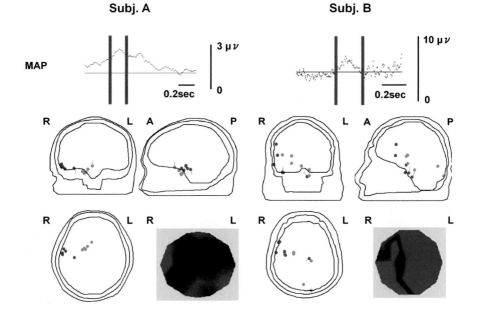

Part IV. (by Homma et al.) **Fig. 3** (P. 131).
Mean absolute potentials (MAP) of all channels and dipole locations (*lower*) were shown in subj. A and B. Dipole locations were calculated between two vertical bars in the pre-insp potentials. Two dipoles, paired with red and green dots, were simultaneously estimated and shown in the 3 dimensional slices (coronal, saggital and axial). A topography of the pre-insp potentials was shown in the right bottom of each subject.

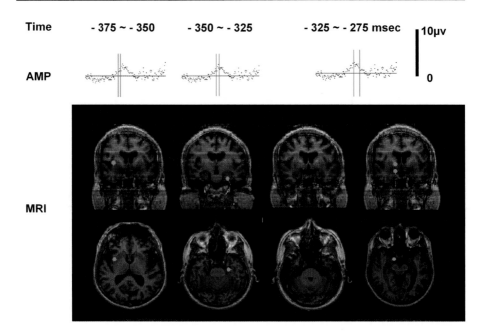

Part IV. (by Homma et al.) **Fig. 4** (P. 132).
The mean absolute potential (MAP) and dipole locations, superimposed on the MRI slices, during the pre-insp potential in subj. A. Dipole locations between 375 to 350 msec (*left*), between 350 to 325 msec (*middle*) and between 325 to 275 msec (*right*) before the onset of inspiration.

Part I
Central Regulation of Breathing

Visualization of Respiratory Neuron Activity in the Ventral Medulla from a Newborn Rodent

Hiroshi Onimaru[1], Akiko Arata[2], Satoru Arata[3], and Ikuo Homma[1]

[1]Department of Physiology, Showa University School of Medicine, Tokyo 142-8555, Japan
[2]Lab. for Memory and Learning, Brain Science Institute, RIKEN, Wako, Saitama 351-0198, Japan
[3]Center for Biotechnology, Showa University, Tokyo 142-8555, Japan

Summary. We visualized respiratory neuron network activity in the medulla of the rat and mouse in vitro by optical recordings using voltage-sensitive dye. The brainstem and spinal cord of 0- to 1-day-old Wistar rats and 0-day-old mice isolated under deep ether anaesthesia were incubated in a modified Krebs solution containing a fluorescent voltage-sensitive dye. Fluorescence signals corresponding to respiratory activity were detected by a CCD image sensor. Pre-inspiratory neuron activity appeared in the limited region of the rostral ventrolateral medulla [i.e. para-facial respiratory group (pFRG) region], preceding the onset of inspiratory activity by about 500 ms. During the inspiratory phase, plateau activity appeared in the more caudal ventrolateral medulla at the level of most rostral roots of the XIIth nerve (i.e. the pre-Bötzinger complex level). We found that pre-inspiratory neurons which were a predominant subtype of the pFRG neurons were located in the area immediately beneath the ventral pia mater at the level of the facial nucleus. We also analyzed respiratory neuron activity in the wild type and two kinds of knock-out mice that exhibit respiratory failure leading to neonatal death due to dysfunction of central respiratory neuron activity. The optical recordings clearly detected the difference in the spatio-temporal pattern between the wild type and both knockout mice.

Key words. Respiratory neurons, optical recordings, ventrolateral medulla, rat, mouse

1 Introduction

The neural circuit that generates the respiratory rhythm in mammals is located in the medulla of the lower brainstem. The main groups of medullary respiratory neurons are thought to be distributed rostrocaudally in the ventrolateral parts of the medulla, where they play a role in respiratory rhythm and inspiratory pattern generation (Feldman, 1986; Bianchi et al., 1995; Ballanyi et al., 1999). To understand the macroscopic behavior of the rostrocaudally-extending respiratory neuron network in the medulla, we stained a brainstem-spinal cord preparation isolated from newborn rats with a voltage-sensitive dye and then applied optical recordings. The first report of this type of optical recording for respiratory neuron activity appeared in Onimaru et al. (1996) in which inspiratory-like activity was detected in the cut surface of the coronary section of the medullary block preparation. Optical detection of respiratory neuron network activity from the ventral surface was reported by Tokumasu et al. (2001). More recently, we performed a detailed analysis of the spatio-temporal pattern using data acquisition by summation of optical signals from a certain pre-triggering point (Onimaru and Homma, 2003). Here, we briefly review our recent findings based on optical recordings in the in vitro preparation from newborn rats and mice.

2 Preparations and Optical Recording Methods

Wistar rats (0-1 days old) and newborn mice (0 day old) were deeply anesthetized with ether until the nociceptive reflexes were abolished. The cerebrum was quickly removed by transection at the intercollicular level; the cerebellum was also removed, and the brainstem and spinal cord were isolated according to methods described previously (Suzue, 1984). For the optical recordings, the preparation was superfused continuously at 2.5-3 ml/min in a 1-ml chamber with modified Krebs solution composed of NaCl, 124 mM; KCl, 5.0 mM; KH_2PO_4, 1.2 mM; $CaCl_2$, 2.4 mM; $MgCl_2$, 1.3 mM; $NaHCO_3$, 26 mM; and glucose, 30 mM. It was equilibrated with 95% O_2 and 5% CO_2 at pH 7.4 and maintained at 26-27°C. Inspiratory motoneuron activity was monitored at the IVth cervical (C4) ventral root with a glass capillary suction electrode.

Detailed procedures for the optical measurements are described elsewhere (Onimaru and Homma, 2003). In brief, the brainstem-spinal cord preparation was incubated in a modified Krebs solution (described above) containing a fluorescent voltage-sensitive dye (25-50 µg/ml Di-2-ANEPEQ for 40-60 min, or 100 µg/ml Di-4-ANEPPS for 35-60 min [Molecular Probes, Inc., Eugene, OR, USA]). Di-2-ANEPEQ, a hydrophilic dye, was basically used for observation from the brainstem surface due to its property of higher diffusion into the tissue than Di-4-ANEPPS, a lipophilic dye. The latter was used for observation from the cut surface of the transverse section due to more stable staining and less noise than the former. After staining, the preparation was placed in a perfusion chamber mounted on a fluorescence microscope stage (BX50WIF-2, Olympus, Tokyo, Japan).

Neuronal activity in the preparation was detected as a change in fluorescence of the voltage-sensitive dye by an optical recording apparatus (MiCAM01, Brain Vision Inc., Tsukuba, Japan) with a charge-coupled device (CCD)-based camera head consisting of 180 x 120 pixels. C4 inspiratory activity was used as the trigger signal for optical recordings, which were performed with an acquisition time (i.e. sampling clock) of 20 msec in most experiments. Fluorescence signals during the respiratory cycles, including the pre-inspiratory phase, were averaged for 40-50 trials.

3 Optical Images in Rat

3.1 The Ventral View

The optical records obtained by the ventral approach showed neuronal activity that preceded inspiratory activity by 400-600 ms and started in the limited region of the rostral ventrolateral medulla; para-facial region, 1.3-1.6 mm lateral to the midline and extended rostrally from the level of the Xth cranial nerve roots (Onimaru and Homma, 2003). Activity then spread caudally and medially in the ventrolateral medulla and some activity was also noted more rostrally during the pre-inspiratory phase. During the inspiratory phase, plateau activity appeared in the more caudal ventrolateral medulla at the level of the most rostral roots of the XIIth cranial nerve (pre-Bötzinger complex level, Smith et al., 1991) and in the ventral horns of the cervical cord.

Pre-inspiratory neuron activity in the para-facial region was further confirmed by optical recordings from the cut surface of the rostral medulla (Onimaru and Homma, 2003). The optical image showed that neuron activity started at the limited region ventrolateral to the facial nucleus and close to the ventral surface during the pre-inspiratory phase and then propagated medially and dorsally during the inspiratory phase. These observations suggest that the major cluster of neurons active during the pre-inspiratory phase is distributed close to the ventral surface such that the optical signals are detectable by the ventral approach.

To confirm the detailed location of Pre-I neurons in the para-facial region ventrolateral to the facial nucleus, we marked the Pre-I neurons with Lucifer yellow and plotted them after whole-cell recordings (Onimaru and Homma, 2003). We found that Pre-I neurons were located in the area immediately beneath the ventral pia mater. These neurons had fusiform or spherical cell bodies with dendrites extending medially and laterally along the ventral surface. Most neurons responded with membrane depolarization to substance P (SP) or 8% CO_2 application in TTX. The arrangement of these Pre-I neurons located close to the ventral surface resembled superficial neurons recently identified in the retrotrapezoid nucleus (RTN) as chemosensitive-glutaminergic neurons (Mulkey et al., 2004).

3.2 The Caudal Cut Surface

In the ventral approach, optical signals during the inspiratory phase were detected mainly in the caudal medulla at the level of the most rostral roots of the XIIth cranial nerves. We speculated that this activity reflects, at least partly, activity in the pre-Bötzinger complex (Reckling and Feldman, 1998; Smith et al., 1991). However, it was also questioned whether optical signals from a deep region such as the pre-Bötzinger complex are detectable by the ventral approach. Therefore, we performed optical recordings from the cut surface of the caudal medulla at the level of the medial roots of the XIIth nerves. Inspiratory activity was detected ventrally to the nucleus ambiguus and in the hypoglossal nucleus (Fig. 1).

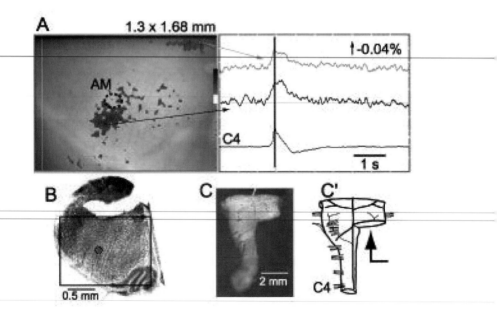

Fig. 1. Optical imagings of respiratory neuron activity in the caudal cut surface of

the medulla. **A**, *left*: Optical image of respiratory neuron activity near the C4 peak (*at the dark vertical line in the right column*) superimposed on the cut surface of the medulla. *Dotted circle* denotes the nucleus ambiguus (AM). *Right*: Fluorescence changes at two different points indicated on A; in the ventrolateral medulla and in the hypoglossal nucleus. Fluorescence decrease, i.e., depolarization is upward. *C4*, electrical record of C4 inspiratory activity. **B**: Approximate level of the optical recording in A. **C, C'**: The photograph and schematic drawing of the preparation. An arrow denotes the direction of observation for the optical recording

4 Optical Images in Mouse

The spatio-temporal patterns in wild type mice were basically identical to those of neonatal rats. The fluorescence change accompanied by respiratory neuron activity appeared prior to the onset of inspiratory nerve activity in the para-facial region (Onimaru et al., 2004). During the inspiratory phase, activity was detected in multiple sites including an area that overlapped with the facial nucleus or para-facial region, as well as the more caudal ventrolateral medulla at the most rostral level of the XIIth cranial nerve roots (i.e. pre-Bötzinger complex level), and the ventral horn of the spinal cord. Application of a low concentration of SP facilitated recovery from the depressed respiratory rhythm after dye-staining (Fig. 2).

We also analyzed the spatio-temporal pattern of respiratory neuron activity in two kinds of knock-out mice (*Tlx3$^{-/-}$*, *Pbx3$^{-/-}$*) that exhibit respiratory failure leading to neonatal death from dysfunction of central respiratory neuron activity (Rhee et al., 2004; Shirasawa et al., 2000). In preparations of *Tlx3$^{-/-}$* mice, the fluorescence change corresponding to respiratory neuron activity appeared to be more dispersed and weaker than in wild type mice, whereas the optical signals appeared first in the rostral, then in the caudal ventrolateral medulla. These optical recording data are consistent with the findings of previous electrophysiological studies suggesting that the excitatory mutual connections between respiratory neurons were weakened in *Tlx3$^{-/-}$* mice (Shirasawa et al., 2000; Cheng et al., 2004). In preparations of *Pbx3$^{-/-}$* mice, optical signal transition in the rostral to caudal direction in the medulla was not clearly detected. Application of 10 nM SP enhanced inspiratory activity in the ventrolateral medulla, whereas preinspiratory activity in the para-facial region was not clearly detected. These optical imagings were considered to reflect a functional disorder of the respiratory neuron network activity in *Pbx3$^{-/-}$* mice that may be more severe than *Tlx3$^{-/-}$* mice (Onimaru et al., 2004).

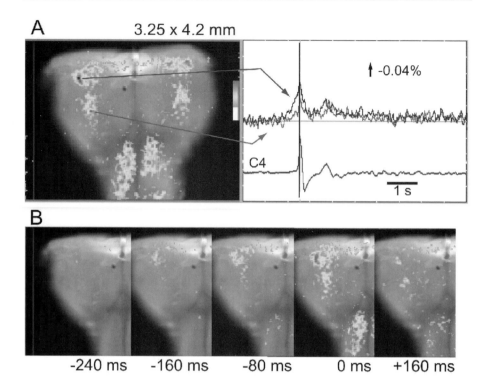

Fig. 2. Optical imagings of respiratory neuron activity in the ventral medulla of a wild type mouse in the presence of 10 nM SP. **A**, *left*: Optical image of respiratory neuron activity near the C4 peak (at the dark vertical line in the right column) superimposed on the ventral surface of the medulla. *Right*: Fluorescence changes at two different points indicated on A; in the rostral ventrolateral medulla at the level just rostral to the IX/Xth cranial roots and in the more caudal ventrolateral medulla at the level of the rostral roots of the XIIth cranial nerve. Fluorescence decrease, i.e., depolarization is upward. *C4*, electrical record of C4 inspiratory activity. **B**: Image is superimposed on the ventral surface of the right half of the medulla. The numbers below each image denote the time from the peak of C4 inspiratory activity. Refer to color plates

5 The Para-Facial Respiratory Group and its Function

The optical recording studies revealed that the respiratory activity in the rostral ventral medulla extends more rostrally and laterally than previously known from electrophysiological studies (Onimaru et al., 1987; Arata et al., 1990). Although Pre-I neurons were predominant respiratory neurons in this area, inspiratory and expiratory neurons were also located. We proposed that localization of this rostral respiratory group could be figured out in positional relation to the facial nucleus; ventral, lateral and caudal to this nucleus. Thus, we referred to this respiratory neuron group as the 'para-facial respiratory group (pFRG)' (Onimaru and Homma, 2003). The medial part of the pFRG may overlap the RTN, which has been identified as an area where neurons with projections to the ventral respiratory group (VRG) are found (Smith et al., 1989; Ellenberger and Feldman, 1990). The major population of the pFRG seems to be lateral to the RTN. The caudal part of the pFRG overlaps the most rostral part of the ventral respiratory group (Bötzinger complex), i.e., the ventral part of the retrofacial nucleus near the caudal end of the facial nucleus. Partial bilateral lesioning of the pFRG area caused significant reduction in the respiratory rate together with changes in the spatio-temporal pattern of the respiratory neuron activity (Onimaru and Homma, 2003). This supports the notion that the pFRG plays an important role in respiratory rhythm generation.

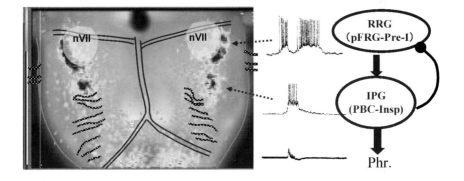

Fig. 3. Optical imaging of respiratory neuron activity in the ventral medulla and hypothetical model of respiratory rhythm generation. The optical imaging was obtained from a 0-day-old rat preparation that was stained with a voltage sensitive dye. Optical imagings from different frames that indicated peak pre-inspiratory activity and peak inspiratory activity were overlapped in one image. The respiratory rhythm generator (*RRG*) is composed of Pre-I neurons in the rostral ventrolateral medulla, including the para-facial respiratory group (*pFRG*). RRG periodically triggers the inspiratory pattern generator (*IPG*), which is composed of inspiratory neurons in the more caudal ventrolateral medulla, including the pre-Bötzinger complex (*PBC*). Membrane potential trajectories are from a Pre-I neuron (*upper*) and from an inspiratory neuron (*middle*), and the bottom trace is C4 inspiratory activity. *nVII*, facial nucleus. *Phr.*, phrenic nerve. Refer to color plates

References

Arata A, Onimaru H, Homma I (1990) Respiration-related neurons in the ventral medulla of newborn rats in vitro. Brain Res Bull 24:599-604

Ballanyi K, Onimaru H, Homma I (1999) Respiratory network function in the isolated brainstem-spinal cord of newborn rats. Prog Neurobiol 59:583-634

Bianchi AL, Denavit-Saubie M, Champagnat J (1995) Central control of breathing in mammals: neuronal circuitry, membrane properties, and neurotransmitters. Physiol Rev 75:1-45

Cheng L, Arata A, Mizuguchi R, Qian Y, Karunaratne A, Gray PA, Arata S, Shirasawa S, Bouchard M, Luo P, Chen CL, Busslinger M, Goulding M, Onimaru H, Ma Q (2004) Tlx3 and Tlx1 are post-mitotic selector genes determining glutamatergic over GABAergic cell fates. Nat Neurosci 7: 510-517

Ellenberger HH, Feldman JL (1990) Brainstem connections of the rostral ventral respiratory group of the rat. Brain Res 513:35-42

Feldman JL (1986) Neurophysiology of breathing in mammals. In: Handbook of physiology. The nervous system. Intrinsic regulatory systems of the brain. Sect 1 (Bloom FE ed), Bethesda, MD: Am Physiol Soc., pp 463-524

Mulkey DK, Stornetta RL, Weston MC, Simmons JR, Parker A, Bayliss DA, Guyenet PG (2004). Respiratory control by ventral surface chemoreceptor neurons in rats. Nature Neurosci 7:1360-1369

Onimaru H, Arata A, Homma I (1987) Localization of respiratory rhythm-generating neurons in the medulla of brainstem-spinal cord preparations from newborn rats. Neurosci Lett 78:151-155

Onimaru H, Kanamaru A, Homma I (1996) Optical imaging of respiratory burst activity in newborn rat medullary block preparations. Neurosci Res 25:183-190

Onimaru H, Homma I (2003) A novel functional neuron group for respiratory rhythm generation in the ventral medulla. J Neurosci 23:1478-1486

Onimaru H, Arata A, Arata S, Cleary ML (2004) In vitro visualization of respiratory neuron activity in the newborn mouse ventral medulla. Dev Brain Res 153: 275-279

Rekling JC, Feldman JL (1998) PreBötzinger complex and pacemaker neurons: hypothesized site and kernel for respiratory rhythm generation. Annu Rev Physiol 60:385-405

Rhee JW, Arata A, Selleri L, Jacobs Y, Arata S, Onimaru H, Cleary ML (2004) Pbx3 deficiency results in central hypoventilation. Amer J Pathol 165:1343-1350

Shirasawa S, Arata A, Onimaru H, Roth KA, Brown GA, Horning S, Arata S, Okumura K, Sasazuki T, Korsmeyer SJ (2000) Rnx (Hox11L2)-deficiency results in congenital central hypoventilation. Nature Gen 24: 287-290

Smith JC, Morrison DE, Ellenberger HH, Otto MR, Feldman JL (1989) Brainstem projections to the major respiratory neuron populations in the medulla of the cat. J Comp Neurol 281:69-96

Smith JC, Ellenberger HH, Ballanyi K, Richter DW, Feldman JL (1991) Pre-

Bötzinger complex: a brainstem region that may generate respiratory rhythm in mammals. Science 254:726-729

Suzue T (1984) Respiratory rhythm generation in the in vitro brain stem-spinal cord preparation of the neonatal rat. J Physiol 354:173-183

Tokumasu M, Nakazono Y, Ide H, Akagawa K, Onimaru H (2001) Optical recording of spontaneous respiratory neuron activity in the rat brain stem. Jpn J Physiol 51:613-619

Search for Genes Involved Central Respiratory Failure Using Mice Models

Satoru Arata[1], Hiroshi Onimaru[2], Akiko Arata[2, 3], Ikuo Homma[2], Tomio Inoue[4], Seiji Shioda[5], Seiji Shirasawa[6], and Michael L. Cleary[7]

[1]Center for Biotechnology, Showa University, 1-5-8 Hatanodai, Shinagawa-ku, Tokyo 142-8555, Japan
[2]Department of Physiology, Showa University School of Medicine, 1-5-8 Hatanodai, Shinagawa-ku, Tokyo 142-8555, Japan
[3]Laboratory for Memory and Learning, Brain Science Institute, RIKEN, Wako, Saitama 351-0198, Japan
[4]Department of Physiology, Showa University School of Dentistry, 1-5-8 Hatanodai, Shinagawa-ku, Tokyo 142-8555, Japan
[5]Department of Anatomy, Showa University School of Medicine, 1-5-8 Hatanodai, Shinagawa-ku, Tokyo 142-8555, Japan
[6]Department of Pathology, International Medical Center of Japan, Shinjuku-ku, Tokyo 162-8655, Japan
[7]Department of Pathology, Stanford University School of Medicine, Stanford, CA 94305, USA

Summary. To search for genes that cause human congenital respiratory failure, we undertook several studies using wild type mice and gene-deficient mice that exhibit respiratory failure. 1) To assess respiratory abnormality, ventilatory parameters were determined by whole-body plethysmography of prenatal and neonatal mice. 2) To clarify the mechanism of respiratory failure, respiratory neuron activity was examined by electrophysiological and optical imaging analysis using a brainstem-spinal cord preparation. 3) To search for the genes related to central respiratory failure in brainstem, alteration of gene expression in the medulla oblongata during transition from embryonic respiration to external respiration was examined by DNA microarray. To date, we have studied several lines of gene-deficient mice exhibit respiratory failure, and we propose a model of central respiratory failure as a syndrome resembling congenital central hypoventilation.

Key words. Central respiratory failure, brainstem-preparation, gene-deficient mice

1 The Assessment of a Central Respiratory Abnormality

Although gene-deficient mice that exhibit respiratory failure die during delivery or within 24 hours of birth, some of the pups look normal and have apparently normal behavior until just before death. So, to uncover the cause of respiratory failure, embryonic mice (E18.5) and newborn mice were examined by whole body plethysmography. Ventilation parameters such as respiratory duration, amplitude, and respiratory rate were examined by this measurement. With wildtype E18.5 mice obtained by cesarean section, measurement of E18.5 mice was performed ~2 to 4 h after external respiration had begun, when the normal pattern had been established. Next, electro-physiological analyses using brainstem-spinal cord preparations from newborn mice or E18.5 mice were used to look at functional disorders in the respiratory center (Arata et al. 1990). In brief, respiratory-like activity corresponding to the inspiratory rhythm was monitored at the C4 ventral root through a glass capillary suction electrode, and membrane potentials of inspiratory neurons and pre-inspiratory (Pre-I) neurons in the ventrolateral medulla were recorded using conventional whole-cell patch clamp methods. On the nerve circuit of the wildtype mouse respiratory center, a coordinate pattern exists among the Pre-I neurons, I neurons, and C4 motoneuron output. This coordinate pattern was also examined by optical recording using voltage-dependent dye (Onimaru et al. 2004).

2 Study of Gene-Deficient Mice that Exhibit Central Respiratory Failure

Tlx3 (previously known as Rnx and Hox11L2) is one of the orphan homeobox genes. Tlx3 deficient mice (Tlx3$^{-/-}$) exhibit respiratory dysfunction and all die within 24 hours of birth. Our previous studies demonstrated that the respiratory pattern in Tlx3$^{-/-}$ mice is characterized by a high respiratory rate with short inspiratory duration and frequent apnea (Shirasawa et al. 2000). Since the short burst duration in Tlx3$^{-/-}$ mice might be related to abnormality of GABA-dependent inhibition, here we examined the effects of a GABAA-antagonist, bicuculline, on burst pattern and respiratory activity. Application of 2 μM bicuculline does not induce significant change in either C4 duration or burst rate of wildtype mice. In contrast, the C4 duration in Tlx3$^{-/-}$ mice significantly increases and the burst rate significantly decreases after 2μM bicuculline treatment to levels comparable to what is seen in wildtype mice (Table 1). Similar changes are induced by 10 μM

Table 1. Summary of *in vivo* and *in vitro* respiratory behavior of gene-deficient mice

	Tlx3[-/-] mice	Pbx3[-/-] mice
Phenotype (*in vivo*)		
Death	within 24 h of birth	within 2 h of birth
Crying	yes	no
Cyanosis	yes	yes
Apnea	commonly occurs	rarely occurs
Respiratory failure (*in vitro*)		
Disorder	central hypoventilation	central hypoventilation
Rate	tachypnea and apnea	irregular
Duration	shorter than wildtype	no change
Neuron network	exists	exists
Bicuculline affects	yes	no
Optical imaging (*in vitro*)		
Neuronal activity	dispersed and weaker	markedly dispersed and weaker
Signal transition	yes	not clear

picrotoxin, a chloride channel blocker, and with a low Cl⁻ solution. Since bicuculline causes transformation of the burst pattern to a quasi-normal pattern, shunting inhibition by excessive or persistent activation of GABAA-dependent Cl⁻ channels of respiratory neurons in Tlx3[-/-] mice may be a mechanism of short burst activity. Furthermore, immunohistochemical studies using Tlx3[-/-] and/or its family member Tlx1[-/-] mice revealed that both genes function as selector genes to promote glutamatergic over GABAergic differentiation during neuronal development (Cheng et. al. 2004). These results suggest that respiratory failure in Tlx3[-/-] mice is caused by excess GABA mediated inhibition.

Pbx3, a member of a TALE (three amino acid loop extension) class homeodomain transcription factor family, is expressed predominantly in the developing central nervous system, including in a region of the medulla oblongata. This protein co-expresses with Tlx3 in the ventrolateral medulla and is capable of a forming a DNA-binding hetero-oligomeric complex with Tlx3. Furthermore, Pbx3[-/-] mice develop to term but die within a few hours of birth, similar to Tlx3[-/-] mice. These data suggest that downstream pathways regulated by hetero-oligomeric Pbx3/Tlx3 transcriptional complexes may be critical for the respiratory network (Rhee et al, 2004). Thus, a comparison of *in vivo* and *in vitro* respiratory activity in Pbx3[-/-] mice and Tlx3[-/-] mice was performed. As summarized in Table 1, Pbx3[-/-] mice exhibit central hypoventilation with an irregular respiratory rate and small burst activity. However, some differences in respiratory activity between the mutant mice were detected, including in frequency of apnea and respiratory duration. Furthermore, the application of bicuculline does not have the transformative effect on the burst pattern of Pbx3[-/-] mice that it does on Tlx3$^{-/-}$ mice. Finally, optical recordings used to look at respiratory activity indicate that a functional disorder of respiratory neuron network activity may be more severe in Pbx3$^{-/-}$ mice than in Tlx3$^{-/-}$ mice (Onimaru et al. 2004). Taken together, these results suggest that the functional disorder of the respiratory center is different in Pbx3$^{-/-}$ mice than in Tlx3$^{-/-}$ mice.

As these data indicate, the mechanisms of central respiratory disorders are complex and our understanding of the mechanisms could benefit from identification of downstream genes and other genes with roles in respiratory failure.

3 Search for Genes Related to Central Respiratory Failure

To screen for genes related to central respiratory failure, gene expression in the medulla oblongata of E18.5 wildtype pups before (0 hour) and after the beginning of external respiration were compared. Some candidate genes were obtained. For example, the expression of the aryl hydrocarbon receptor nuclear translocator 2 gene (ARNT2) increases after external respiration. In addition, ARNT2 deficient mice have already been generated, and the mutant mice die within 24 hours of birth (Hosoya et al. 2001). Our current focus is on the study of mouse models of respiratory failure and on identification of additional candidate genes.

References

Arata A, Onimaru H, Homma (1990) Respiration-related neurons in the ventral medulla of newborn rats in vitro. Brain Res Bull 24:599-604

Cheng L, Arata A, Mizuguchi R, Qian Y, Karunaratne A, Gray PA, Arata S, Shirasawa S, Bouchard M, Luo P, Chen CL, Busslinger M, Goulding M, Onimaru H, Ma Q (2004) Tlx3 and Tlx1 are post-mitotic selector genes determining glutamatergic over GABAergic cell fates. Nat Neurosci 7 (5): 510-7

Hosoya T, Oda Y, Takahashi S, Morita M, Kawauchi S, Ema M, Yamamoto M, Fujii-Kuriyama Y (2001) Defective development of secretory neurons in the hypothalamus of Arnt2-knockout mice. Genes to Cells 6:361-374

Onimaru H, Arata A, Arata S, Shirasawa S, Cleary ML (2004) In vitro visualization of respiratory neuron activity in the newborn mouse ventral medulla. Brain Res Dev Brain Res 153:275-9

Rhee JW, Arata A, Selleri L, Jacobs Y, Arata S, Onimaru H, Cleary ML (2004) Pbx3 deficiency results in central hypoventilation. Am J Pathol 165(4): 1343-50

Shirasawa S, Arata A, Onimaru H, Roth KA, Brown GA, Horning S, Arata S,

Okumura K, Sasazuki T, Korsmeyer SJ (2000) Rnx deficiency results in congenital central hypoventilation. Nat Genet. 24:287-90

Control of Central Histaminergic Neurons for Respiration during Hypercapnia in Conscious Mice

Michiko Iwase, Masahiko Izumizaki, Kenichi Miyamoto*,
Mitsuko Kanamaru, and Ikuo Homma

2nd Department of Physiology, Showa University School of Medicine, 1-5-8 Hatanodai, Shinagawa-ku, Tokyo 142-8555, Japan
*Present address: 2nd Department of Internal Medicine, Nara Medical University, Kashihara 634-8521, Japan

Summary. Central histaminergic neurons located in the posterior hypothalamus affect many autonomic behaviors. We examined the roles of these neurons in respiration in conscious histamine H1 receptor-knockout (H1RKO) and wild-type (WT) mice. Acute stepwise hypercapnia increased respiratory frequency (f), tidal volume (V_T), and minute ventilation (V_E) in both genotypes. However, H1RKO mice showed a lower f response and a higher VT response than those of WT mice. The V_T -T_I relation curve for H1RKO mice was shifted to the right and upward relative to that for WT mice, suggesting that in H1RKO mice the termination of inspiration is delayed by an increase in the inspiratory off switch (IOS) threshold. Increased body temperature increased the f response to hypercapnia in WT but not in H1RKO mice. This polypnea during hypercapnia was due mainly to a reduction in T_E. These results suggest that central histaminergic neurons contribute to the termination of inspiration in the IOS mechanism and to polypnea during hyperthermia via activation of H1 receptors.

Key words. Central histaminergic neuron, respiration, inspiratory off switch, hyperthermia, CO_2 response

1 Introduction

Respiratory pattern, generated in the lower brainstem, is modified by inputs from behavioral and metabolic control of respiration. Histaminergic neurons are localized in the posterior hypothalamus and project axons to the lower brainstem, limbic system, and cerebral cortex (Inagaki et al., 1988; Iwase et al., 1993), and they receive inputs from the cerebral limbic system (Ericson et al., 1991). Central histaminergic neurons modify behavioral activities such as the sleep-awake cycle, locomotor activity, learning and memory, feeding and drinking, and thermoregulation through H1, H2, and H3 receptors. Based on results of morphologic and physiologic studies, we hypothesized that histaminergic neurons modulate respiratory pattern generation through the behavioral control of respiration.

Hypothalamic H1 receptors activate lipolysis in adipose tissue by activating the sympathetic nervous system (Tsuda et al., 2002). Therefore, alteration of metabolism influences respiration via central and peripheral chemoreceptors, central histaminergic neurons potentially affect the metabolic control of respiration. We hypothesized that histaminergic neurons also modulate respiratory pattern generation through the metabolic control of respiration.

To clarify the role of central histaminergic neurons in the behavioral and metabolic control of respiration, we examined respiratory pattern during the hypercapnic ventilatory response and the effect of body temperature (BT) on this responses in conscious histamine H1 receptor- knock out (H1RKO) and wild-type (WT) mice (Izumizaki et. al., 2000b; Miyamoto et al., 2004). We described the role of histaminergic neurons in the respiratory control during hypercapnia on the bases of previous our studies.

2 Methods

2.1 Mice

H1RKO mice were provided by Dr. T. Watanabe (Medical Institute of Bioregulation, Kyusyu University, Fukuoka, Japan) and were bred at Showa University under specific pathogen-free conditions. Tail biopsies were analyzed for presence of the H1 receptor mutant allele by polymerase chain reaction of genomic DNAs as described previously (Inoue et al., 1996). H1RKO and C57BL/6 mice were backcrossed, and male littermates were used in some experiments. Mice were provided with food and water *ad li-*

bitum, housed at a controlled temperature (24°C), and exposed to a daily 12-h light-dark cycle. All experiments were conducted at 24°C between 10:00 a.m. and 5:00 p.m. The study protocol was approved by the Showa University Animal Experiments Committee.

2.2 Measurement and Control of BT

Measurement and control of BT was carried out as described previously (Izumizaki et al., 2000a). In brief, mice were anesthetized with pentobarbital for placement into a body plethysmograph. BT was measured by a thermistor probe inserted into the rectum. BT at 37-38°C was regarded as normothermia. To induce hyperthermia, mice were warmed to 39°C with a heat lamp from outside the plethysmograph body chamber.

2.3 Measurement of Lung Ventilation

To study respiratory pattern generation during hypercapnia, anaesthetized and unrestrained mice were placed in a whole-body plethysmograph (PLY3211; Buxco Electronics, Sharon, CT, USA) as described previously (Miyamoto et al., 2004). To study BT, mice were placed in a double-chamber plethysmograph, and a thermistor probe was inserted into the rectum to monitor BT as described previously (Izumizaki et al., 2000a, b). After exposure to 100% O_2, the inhalation gas was switched stepwise to 5%, 7%, and 9% CO_2 in O_2 at 5-min intervals. Tidal volume (V_T, ml BTPS), inspiratory time (T_I, s), expiratory time (T_E, s), frequency (f, breaths/min) and minute ventilation ($\dot{V}_E$, ml BTPS) were measured. V_T and $\dot{V}_E$ were normalized per 10 g body weight.

2.4 Analysis of Blood Gases

We measured $PaCO_2$ directly during hypercapnic gas exposure in conscious mice. Mice were supplied with room air and gases composed of 5% CO_2 and 9% CO_2 balanced with O_2. Insertion of an arterial catheter was carried out as described previously (Iwase et al., 2004).

2.5 Statistical Analysis

Data are shown as mean ± SEM values. Two-way repeated-measures analysis of variance (ANOVA), analysis of covariance (ANCOVA), and

Student's t-test were performed with a commercially available software package (SPSS, SPSS Japan Inc., Tokyo, Japan). Statistical significance was set at $P < 0.05$.

3 Results

3.1 Changes in Respiratory Pattern during Hypercapnic Ventilatory Responses

The f, V_T, and $\dot{V}_E$ were plotted against $PaCO_2$ during 5% and 9% CO_2 exposure. During hypercapnia, H1RKO mice showed a lower f response and a higher V_T response than those of WT mice. However, there was no significant difference in $\dot{V}_E$ between genotypes (Fig. 1a,b,c). The lower f response in H1RKO mice was accompanied by increases in both T_I and T_E. The V_T - T_I curves for the two genotypes during stepwise CO_2 exposure, regressed to the hyperbola as described by Clark and von Euler (1972) (Fig. 1d). CO_2 inhalation increased V_T and reduced T_I in both genotypes. The hyperbolic curve for H1RKO mice was shifted to the right and upward relative to that for WT mice.

3.2 Effects of Increased BT on Respiratory Variables during Hypercapnia

Respiratory variables in response to inspired CO_2 were examined at normal BT and at 39°C in WT and H1RKO mice. CO_2 produced significant increases in f, V_T and $\dot{V}_E$ at both BTs in both genotypes. There was a significant increase in f and a decrease in V_T at 39°C in WT mice. Changes in BT had no effect on $\dot{V}_E$. The increase in f at 39°C did not occur in H1RKO mice (Fig. 2). In WT mice, BT-induced polypnea was caused by reductions in both T_I and T_E. However, the reduction in T_E was more significant for polypnea at 39°C. In H1RKO mice, there were no differences in T_I, T_E or f between BTs.

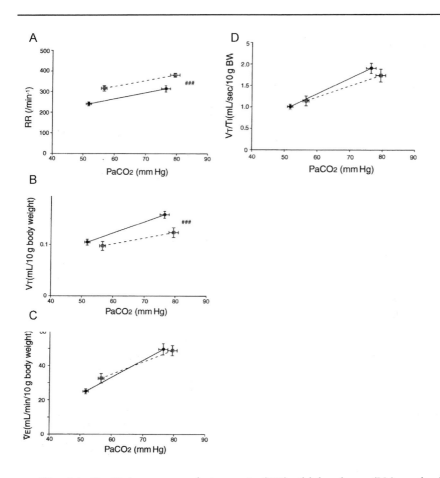

Fig. 1A, B, C Average respiratory rate (RR), tidal volume (V_T), and minute ventilation ($\dot{V}_E$) during each 5% and 9% CO_2 exposure plotted against mean $PaCO_2$. H1 receptor-knockout mice (filled circles) show a significantly smaller RR response (###$P < 0.001$) and larger VT response (###$P < 0.001$) to hypercapnia than responses of wild-type mice (open squares). Data were tested by ANCOVA. [From Miyamoto et al., 2004.] **D** Average tidal volume (V_T) during successive 3, 5, 7, and 9% CO_2 exposures plotted against the corresponding inspiratory time (T_I) for wild-type (open squares) and H1 receptor-knockout mice (filled circles).[From Miyamoto et al., 2004.]

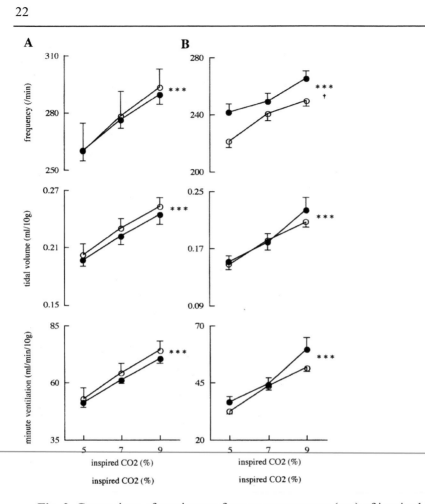

Fig. 2. Comparison of respiratory frequency responses (*top*) of inspired CO_2 at 37-38°C (*filled circles*) and at 39°C (*open circles*), V_T (*middle*) and $\dot{V}_E$ (*bottom*) in wild-type (**A**) and histamine H1 receptor-knockout mice (**B**). There was a significant difference in f between 37-38°C and 39°C (*$P < 0.05$). There was a significant effect with respect to inspired CO_2 (††$P < 0.01$, †††$P < 0.001$). There was a significant effect with respect to temperature (‡$P < 0.05$). [From Izumizaki et al., 2000b.]

4 Discussion

The role of central histaminergic neurons in respiratory pattern formation was examined in H1RKO and WT mice. Acute stepwise hypercapnia increased f, V_T, and $\dot{V}_E$ in both genotypes, and respiratory patterns differed between the genotypes. H1RKO mice showed a lower f response and a higher V_T response than those of WT mice. The V_T-T_I relation curve for H1RKO mice was shifted to the right and upward relative to that for WT mice, indicating that an increased inspiratory off switch (IOS) threshold delays the termination of inspiration. Dutschmann et al. (2003) found that histamine affects inspiratory and expiratory neurons and that histamine is tonically active in the respiratory network in mice. H1 receptors may be involved in the IOS mechanism in the lower brainstem.

Respiratory pattern is also affected by thermal inputs from the hypothalamus. In WT but not in H1RKO mice increased BT increased the f response during hypercapnia, and the reduced T_E accounted for the polypnea. These results are consistent with those of a physiologic study with an inhibitor of histamine synthesis, α-fluoromethylhistidine (α-FMH) (Izumizaki et al., 2000a). Evidence from several studies supports our hypothesis that histamine affects the behavioral control of respiration. Central histamine regulates heat-loss mechanisms in behavioral studies (Green et al., 1976) and causes hypothermia in mice (Chen et al., 1995). There is a distribution of histaminergic neuronal fibers in the preoptic area known as the thermoregulatory center (Inagaki et al., 1988). We previously showed that an H1 receptor antagonist injected into the preoptic area or α-FMH injected into the lateral ventricle reduces BT-induced polypnea (Iwase et al., 2001). Thus, we propose that central histaminergic neurons contribute to BT-induced polypnea through this thermoregulatory center.

The IOS mechanism is affected during thermal polypnea. Because thermal inputs shift the V_T-T_I relation curve below to the left, indicating a lowered IOS threshold. Hyperthermia elicits histamine release in the rostral ventrolateral medulla and the solitary nucleus, where the respiratory neuron network is located (Kanamaru et al, 2001). The released histamine in these areas may be advantageous for thermal polypnea, because central histamine activates the IOS mechanism.

In conclusion, central histaminergic neurons contribute to the termination of inspiration in the IOS mechanism and to polypnea during hyperthermia via activation of H1 receptors.

References

Chen Z, Sugimoto Y, Kamei C (1995) Effects of intracerebroventricular injection of histamine and its related compounds on rectal temperature in mice. Methods Find Exp Clin Pharmacol 17:669-675

Clark FJ, von Euler C (1972) On the regulation of depth and rate of breathing. J Physiol 222:267-295

Dutschmann M, Bischoff AM, Busselberg D, Richter DW (2003) Histaminergic modulation of the intact respiratory network of adult mice. Pflugers Arch 445:570-576

Ericson H, Blomqvist A, Kohler C (1991) Origin of neuronal inputs to the region of the tuberomammillary nucleus of the rat brain. J Comp Neurol 311:45-64

Green MD, Cox B, Lomax P (1976) Sites and mechanisms of action of histamine in the central thermoregulatory pathways of the rat. Neuropharmacology 15:321-324

Inagaki N, Yamatodani A, Ando-Yamamoto M, Tohyama M, WatanabeT, Wada H (1988) Organization of histaminergic fibers in the rat brain. J Comp Neurol 273:283-300

Inoue I, Yanai K, Kitamura D, Taniuchi I, Kobayashi T, Niimura K, Watanabe T, Watanabe T (1996) Impaired locomotor activity and exploratory behavior in mice lacking histamine H1 receptors. Proc Natl Acad Sci USA 93:13316-13320

Iwase M, Homma I, Shioda S, Nakai Y (1993) Histamine immunoreactive neurons in the brain stem of the rabbit. Brain Res Bull 32:267-272

Iwase M, Izumizaki M, Kanamaru M, Homma I (2001) Involvement of central histaminergic neurons in polypnea induced by hyperthermia in rabbits. Neurosci Lett 298:119-122

Iwase M, Izumizaki M, Kanamaru M, Homma I (2004) Effects of hyperthermia on ventilation and metabolism during hypoxia in conscious mice. Jpn J Physiol 54:53-59

Izumizaki M, Iwase M, Kimura H, Kuriyama T, Homma I (2000a) Central histamine contributed to temperature-induced polypnea in mice. J Appl Physiol 89 770-776

Izumizaki M, Iwase M, Kimura H, Yanai K, Watanabe T, Watanabe T, Homma I (2000b) Lack of temperature-induced polypnea in histamine H1 receptor-deficient mice. Neurosci Lett 284:139-142

Kanamaru M, Iwase M, Homma I (2001) Neuronal histamine release elicited by hyperthermia mediates tracheal dilation and pressor response. Am J Physiol Regul Integr Comp Phyisol 280:R1748-R1754

Miyamoto K, Iwase M, Kimura H, Homma I (2004) Central histamine contributes to the inspiratory off-switch mechanism via H1 receptors in mice. Respir Physiol Neurobiol 144:25-33

Tsuda K, Yoshimatsu H, Niijima A, Chiba S, Okeda T, Sakata T (2002) Hypothalamic histamine neurons activate lipolysis in rat adipose tissue. Exp Biol Med 227:208-213

Part II
Neurogeneration and Neuroprotection

The Role of the Blood-Brain Barrier in Feeding: Leptin

William A. Banks[1, 2], Susan A. Farr[1, 2], and John E. Morley[1, 2]

[1]Geriatric Research, Educational, and Clinical Center, Veterans Affairs Medical Center, St. Louis, MO, USA
[2]Division of Geriatrics, Department of Internal Medicine, Saint Louis University School of Medicine, St. Louis, MO, USA

Summary. The blood-brain barrier (BBB) has emerged as a regulatory interface that controls the exchange of informational molecules between the blood and the central nervous system (CNS) fluids (the brain interstitial and the cerebrospinal fluid). As such, it is a pivotal point in a humoral-based, endocrine-like communication between the CNS and peripheral tissues. For example, the BBB controls the entry of major feeding hormones into the CNS from the blood, including leptin. Impaired transport of leptin across the BBB is an early cause of leptin resistance. Leptin transport is regulated by alpha$_1$-adrenergics, which simulate transport, and triglycerides, which inhibit it. The ability of triglycerides to inhibit leptin transport may have evolved as a response to the hypertriglyceridemia of starvation. If so, their ability to inhibit leptin transport in obesity may be a metabolic case of mistaken identity.

Key words. Leptin, blood-brain barrier, obesity, epinephrine, triglycerides

1 Introduction

Evidence for the blood-brain barrier (BBB) dates back to the late 19th century when it was found that peripherally injected dyes and bile acids could not access the brain (Davson, 1967). These substances were found to tightly bind to serum albumin which, in turn, was unable to cross the BBB.

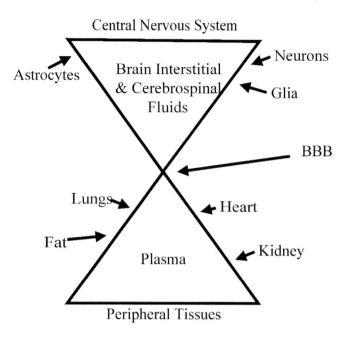

Fig. 1. Schematic of BBB as pivotal point in communication between CNS and peripheral tissues

Besides limiting uncontrolled access of circulating substances into the blood, the BBB also allows on a selective basis many substances into and out of the central nervous system (CNS). As such, the BBB has emerged as a regulatory interface (Banks and Kastin, 1990) that controls the exchange of informational molecules between the blood and the CNS fluids (the brain interstitial fluid and the cerebrospinal fluid). It is a pivotal point in a humoral-based, endocrine-like communication between the CNS and peripheral tissues (Fig. 1).

The BBB of the adult mammal actually consists of several barriers in parallel (Davson and Segal, 1996). The vascular bed of the brain is specially modified in several ways to prevent the unregulated entry of serum proteins into the CNS. The major modifications are: a sealing of the endothelial cells comprising the capillary bed and lining the arterioles and venules by tight junctions, thus preventing intercellular leaks; a diminished number of capillary fenestrea; decreased pinocytotic activity. These last two modifications decrease intracellular leaks. The choroid plexus is comprised of ependymal cells joined by tight junctions and separates the cerebrospinal fluid (CSF) from the circulation. Tanycytic barriers separate the circumventricular organs (small areas of the brain near the ventricles with diminished vascular barriers) from adjacent brain tissue. Together, these barriers produce one of the highest gradients in biology: the CSF/serum ratio for serum albumin is about 1:200.

Beside these physical attributes, the BBB is endowed with other characteristics which may, particularly for some substances, further limit their blood-to-brain transit. Two such characteristics are enzymatic activity and efflux systems (Brownlees and Williams, 1993). The BBB contains many enzymes which can be a major obstacle for limiting the entry of some substances into the brain. For example, enzymatic degradation of angiotensin II by the vascular BBB is a major limiting factor in preventing that peptide from entering the CNS. Efflux systems in the brain can also prevent many substances from entering the brain. For example, the P-glycoprotein system effectively or diminishes the accumulation in brain of many small, lipid soluble molecules.

Any protein or peptide entering the brain must do so by negotiating the BBB. Many saturable transporters located at the BBB are themselves controlled and regulated by substances and pathophysiological events (Banks, 1999). Interactions between the BBB and leptin illustrates each of these properties.

2 Leptin Resistance and the BBB

Leptin is a 16 kDa protein secreted from fat cells. Its isolation in 1994 revolutionized the perception of obesity and body weight control (Zhang et al., 1994). It is transported across the BBB by a saturable process where it interacts with receptors in the arcuate nucleus where it acts as an anorectic and increases thermogenesis. Leptin induces anorexia primarily by inhibiting orexins such as neuropeptide Y and stimulating anorexins such as melanocyte stimulating hormone (MSH). These effects act to decrease fat mass, and so leptin participates in a negative feed back loop regulating fat mass (Friedman and Halaas, 1998).

In outbred rodents with diet-induced obesity and humans, obesity is associated with a resistance to the actions of leptin. Resistance can, in theory, arise from one of three general mechanisms: 1) an inability of serum leptin to cross the BBB, 2) defects in the leptin receptor/post-receptor function, or 3) failure at the downstream neural circuitries (e.g., mutations in MSH). Examples in the failure of each of these three mechanisms have been found in both humans and animals. However the first level of failure occurs at the BBB.

Evidence for the failure of the BBB comes from several sources and shows that the nature of the failure at the BBB has several causes. Caro et al and others have shown that although CSF levels of leptin increase with obesity, this increase is small in comparison to that seen in the serum (Mantyh et al., 1993). For example, in an anorectic individual with a serum leptin level of about 1 ng/ml, the level of leptin in CSF is about 0.1 ng/ml. In an obese person with a serum leptin level of 40 ng/ml, a forty fold increase over the anorectic value, CSF levels are about 0.34 ng/ml, only a 3.4 fold increase. A similar pattern is found in brain in normal body weight mice in which the vascular levels of leptin can be manipulated in a brain perfusion method (Banks et al., 2000). The relative increase in brain and CSF leptin becomes ever smaller as serum leptin levels rise. This means that the relative signal to brain from blood becomes more muted as serum leptin levels increase.

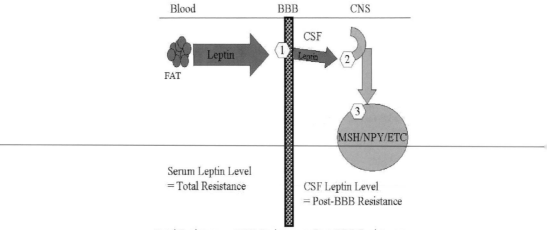

Fig. 2. Resistance to leptin can occur at the BBB (*1*), receptor/post-receptor (*2*), or downstream neuronal circuitries (*3*). Serum levels must overcome resistance at all three levels (total resistance), whereas CSF levels must over come post-BBB resistances (*2* and *3*)

The relative contributions of the BBB and non-BBB mechanisms to leptin resistance can be easily approximated. Increases in blood levels reflect total resistance to leptin (resistance at the BBB, receptor/post-receptor level, and the downstream neural circuitries), whereas CSF levels reflect resistances beyond the BBB (receptor/post-receptor level and the downstream neural circuitries); Fig. 2. For example, total resistance of an obese person with a serum leptin level of 40 ng/ml is 4 times that of a lean person with a serum leptin level of 10 ng/ml: 40/10 = 4. The inverse of 4 is 0.25 and so the system is working at 25% efficiency. Corresponding CSF levels in these two groups are about 0.34 ng/ml and 0.26 ng/ml respectively (Mantyh, Ghilardi, Rogers, DeMaster, Allen, Stimson, and Maggio, 1993), giving a post-BBB resistance of 1.3 (a value of 1.0 would indicate "no resistance") and an efficiency of 76%. Dividing total resistance by post-BBB resistance yields the resistance at the BBB of 3.1 and an efficiency of 32%. This shows that in this particular example, the BBB of the obese person is working at only about 32% of that of the lean person, whereas the post-BBB mechanisms are working at about 76%.

3 Mechanisms of Leptin Transporter Failure

The immediate cause of the decreased efficiency in transport of leptin at the higher serum levels is self saturation. Leptin is transported across the BBB by a saturable transporter (Banks et al., 1996). As serum levels increase, this transporter becomes progressively saturated so that its ability to transport still more leptin becomes reduced. As might be predicted, impaired transport of leptin is acquired with increasing obesity and can be reversed with weight loss (Banks et al., 1999; Banks and Farrell, 2003).

Acquisition of impairment has been shown in two models of diet-induced obesity. The obesity of maturity model takes advantage of the tendency for a subset of the outbred CD-1 strain to gain considerable body weight as it ages, whereas another subset remains at nearly its post-pubertal weight (Pelleymounter et al., 1998). This occurs without placing the CD-1 mice on any special diet and is similar to the age-related weight gain seen in many humans. In those mice which gain weight, leptin transport is impaired, whereas those mice age-matched mice which did not gain weight have a transport rate which is unchanged (Banks, DiPalma, and Farrell, 1999). Furthermore, within a population of CD-1 mice, leptin transport rate across the BBB is inversely related to body weight. In the diet-induced obese (DIO) model, Sprague-Dawley-derived rats have been bred for several generations for the propensity towards the DIO characteristic or for the diet-resistant (DR) characteristic. These rats with a propensity towards DIO are born with a resistance to leptin at the level of the receptor (Levin et al., 2004). At birth and prior to puberty, DIO and DR rats weigh the same and have similar rates of transport across the BBB. However, as DIO rats age, they gain weight relative to DR rats and develop a deficit in BBB transport of leptin. Both weight gain and the defect in BBB transport can be accelerated by placing the DIO rats on a high energy diet.

The BBB deficits in leptin transport as acquired in diet-induced obesity can be reversed, at least in the outbred CD-1 mouse. Weight loss induced by dietary restriction or by treatment with leptin both are associated with an increase in the rate of leptin transport across the BBB (Banks and Farrell, 2003).

These obesity-related alterations in the rate of leptin transport across the BBB could be, in theory, entirely explained by the rise and fall in serum leptin levels. However, causes other than saturation by serum leptin impair transport at the BBB in obesity. Brain perfusion studies show that the BBB of the obese mouse is affected by some factor other than serum leptin (Banks, DiPalma, and Farrell, 1999). The brain perfusion technique replaces blood with buffer and so negates the immediate effects of circulating substances on the BBB. These studies show that the majority of the inhibitory effect of obesity on leptin transport is not caused by the immediate effect of circulating substances, such as leptin. This rules out saturation as the sole explanation for impaired BBB transport. Furthermore, it shows that the leptin transporter is impaired, and possibly regulated, by events or substances related to obesity.

4 Regulation of the Leptin Transporter

Two classes of endogenous substances intimately related to body weight control and obesity have been identified which affect leptin transport. The alpha$_1$ adrenergics and triglycerides have potent and opposing effects on leptin transport across the BBB. Together, these effects on leptin transport explain paradoxic findings and even provide a frame work for explaining leptin resistance itself.

4.1 Epinephrine

Epinephrine through alpha$_1$ adrenergic receptors stimulates leptin transport by 2-3 fold (Banks, 2001). This is mediated thorough a peripheral site either at the luminal surface of the BBB, on the leptin molecule, or elsewhere. The effect is immediately seen when the epinephrine is given intravenously and is long lasting. The effects of a single ip injection of epinephrine lasts for at least 2 h.

Other interactions between epinephrine and leptin are well documented (Fig. 3). Leptin enhances sympathetic nerve activity to fat and other tissues

(Haynes et al., 1997) and stimulates the release of epinephrine into blood by binding to leptin receptors located on the adrenal gland's epinephrine-secreting cells (Cao et al., 1997; Takekoshi et al., 1999; Cao, Considine, and Lynn, 1997). The increase sympathetic tone suppresses leptin secretion from fat cells by binding to β3 adrenergic receptors (Guerre-Millo, 1997; Fritsche et al., 1998). A classic negative feedback loop thus exits in the periphery, explaining the inverse relation between blood levels of epinephrine and leptin (Bottner et al., 1999; Mills et al., 1998). Stimulation of leptin transport into the brain by epinephrine reinforces the ability of epinephrine to lead to reductions in serum leptin levels. A 2 fold increase in the leptin transport rate into brain should lead to an eventual reduction in body fat and leptin levels of about 50%.

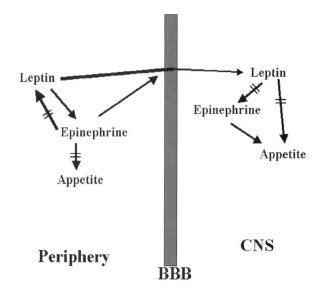

Fig. 3. Relations between peripheral and central epinephrine and leptin in appetite regulation

Epinephrine stimulated leptin transport resolves the contradiction that peripheral epinephrine is anorexic whereas CNS epinephrine is orexigenic with another paradox: whereas leptin promotes adrenergic tone in the periphery, it inhibits the release of epinephrine from neurons (Brunetti et al., 1999). Therefore, serum leptin promotes anorexia but stimulating peripheral sympathetic tone. Increased serum epinephrine promotes entry of leptin into the brain where leptin would inhibit the orexigenic epinephrine release.

4.2 Triglycerides

Triglycerides have also emerged as a major regulator of the transport of leptin across the BBB. Elevating serum triglycerides by starving animals, administering triglycerides, or feeding them a high fat diet inhibits leptin transport, whereas lowering serum triglycerides with short-term fasting or with gemfibrozil stimulates leptin transport (Banks et al., 2004; Kastin and Akerstorm, 2000). Leptin transport is also inhibited when triglycerides are added to an in vitro model of the BBB (Fig. 4). Leptin, in turn, decreases serum triglyceride levels by inducing triglyceride hydrolysis and

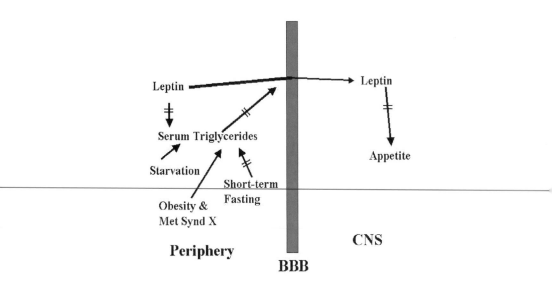

Fig. 4. Relations between triglycerides, leptin, and appetite

FFA oxidation and inhibiting FFA synthesis (Steinberg et al., 2002; Reidy and Weber, 2002). Therefore, the ability of triglycerides to impair leptin action completes a negative feed loop between these two substances.

Elevations in serum triglycerides is an integral part of metabolic syndrome X, which also includes obesity, hypertension, and insulin-resistant diabetes mellitus (Hansen et al., 1999). Inhibition of an anorectic single in the face of obesity seems paradoxical. However, triglycerides are not only elevated in obesity, but also during starvation. If starvation has exerted more influence on evolution than obesity, it may be that elevations in triglycerides have come to be interpreted by the brain as a starvation, rather than an obesity, signal. If so, then inhibition of an anorectic signal to the brain by serum triglycerides makes sense.

Therefore, the effects of triglycerides on the BBB transport of leptin may explain a great deal of the mechanisms underlying the epidemic of obesity. Inhibition of leptin transport across the BBB by triglycerides may have evolved in response to starvation. If so, then the inhibition by triglycerides of the anorectic signaling properties of leptin to brain with obesity may be a metabolic case of mistaken identity.

5 Conclusions

The blood-brain barrier (BBB) plays a crucial role in mediating metabolic signals between the CNS and peripheral tissues. One such example of this is leptin. Leptin as a large molecule would be unlikely to access brain in sufficient quantities if it were not transported across the BBB by a saturable transporter. This transporter fails in obesity and is an earlier level of resistance to leptin than is failure at the leptin receptor. Impairment in leptin transport becomes progressively worse as obesity occurs, setting up a positive or feed-forward cycle between obesity and leptin resistance. Impaired BBB transport of leptin can be reversed with weight loss induced either by dietary modification or leptin treatment. Transport of leptin across the BBB is modulated by epinephrine, which increases leptin transport by 2-3 fold. Triglycerides inhibit leptin transport so that an inverse relation exists between leptin and serum triglyceride levels. Serum triglycerides are elevated not only in obesity but also in starvation and leptin transport is correspondingly inhibited during starvation. Preventing an anorectic like leptin from reaching its receptors in the brain during starvation has clear adaptive value. It may be that the ability of triglycerides to inhibit leptin transport evolved as a response to starvation. In this case, the inhibition seen with obesity is an inappropriate response to triglycerides.

References

Banks WA (1999) Physiology and pathophysiology of the blood-brain barrier: Implications for microbial pathogenesis, drug delivery and neurodegenerative disorders. J. Neurovirology 5:538-555

Banks WA (2001) Enhanced leptin transport across the blood-brain barrier by α1-adrenergic agents. Brain Res. 899:209-217

Banks WA, Clever CM, and Farrell CL (2000) Partial saturation and regional variation in the blood to brain transport of leptin in normal weight mice. Am. J. Physiol. 278:E1158-E1165

Banks WA, Coon AB, Robinson SM, Moinuddin A, Shultz JM, Nakaoke R, and Morley JE (2004) Triglycerides induce leptin resistance at the blood-brain barrier. Diabetes 53:1253-1260

Banks WA, DiPalma CR, and Farrell CL (1999) Impaired transport of leptin across the blood-brain barrier in obesity. Peptides 20:1341-1345

Banks WA and Farrell CL (2003) Impaired transport of leptin across the blood-brain barrier in obesity is acquired and reversible. Am J Physiology 285:E10-E15

Banks WA and Kastin AJ (1990) Editorial Review: Peptide transport systems for opiates across the blood-brain barrier. Am. J. Physiol. 259:E1-E10

Banks WA, Kastin AJ, Huang W, Jaspan JB, and Maness LM (1996) Leptin enters the brain by a saturable system independent of insulin. Peptides 17:305-311

Bottner A, Eisenhofer G, Torpy DJ, Ehrhart-Bornstein M, Keiser HR, Chrousos GP, and Bornstein SR (1999) Preliminary report: Lack of leptin suppression in response to hypersecretion of catecholamines in pheochromocytoma patients. Metab. Clin. Exp. 48:243-245

Brownlees J and Williams CH (1993) Peptidases, peptides, and the mammalian blood-brain barrier. J. Neurochem. 60:793-803

Brunetti L, Michelotto B, Orlando G, and Vacca M (1999) Leptin inhibits norepinephrine and dopamine release from rat hypothalamic neuronal ends. Eur. J. Pharmacol. 372:237-240

Cao GY, Considine RV, and Lynn RB (1997) Leptin receptors in the adrenal medulla of the rat. Am. J. Physiol. 273:E448-E452

Davson H (1967) The blood-brain barrier, in physiology of the Cerebrospinal Fluid pp 82-103, J. and A. Churchill, LTD., London

Davson H and Segal MB (1996) Physiology of the CSF and Blood-Brain Barriers. CRC Press, Boca Raton

Friedman JM and Halaas JL (1998) Leptin and the regulation of body weight in mammals. Nature 395:763-770

Fritsche A, Wahl HG, Metzinger E, Reen W, Kellerer M, Haring H, and Stumvoll M (1998) Evidence for inhibition of leptin secretion by catecholamines in man. Experimental and Clinical Endocrinology and Diabetes 106:415-418

Guerre-Millo M (1997) Regulation of ob gene and overexpression in obesity. Biomedicine and Pharmacotherapy 51:318-323

Hansen, B. C., Saye, J., and Wennogle, L. P. The metabolic Syndrome X: Convergence of insulin resistance, glucose intolerance, hypertension, obesity, and dyslipidemia – Searching for the underlying defects. Hansen, B. C., Saye, J., and Wennogle, L. P. (892), 1-336. 1999. New York, NY, The New York Academy of Sciences. Annals of the New York Academy of Sciences

Haynes WG, Morgan DA, Walsh SA, Mark AL, and Sivitz WI (1997) Receptor-mediated regional sympathetic nerve activation by leptin. J. Clin. Invest. 100:270-278

Kastin AJ and Akerstrom V (2000) Fasting, but not adrenalectomy, reduces transport of leptin into the brain. Peptides 21:679-682

Levin BE, Dunn-Meynell AA, and Banks WA (2004) Obesity-prone rats have normal blood-brain barrier transport but defective central leptin signaling before obesity onset. Am. J. Physiol. 286:R143-R150

Mantyh PW, Ghilardi JR, Rogers S, DeMaster E, Allen CJ, Stimson ER, and Maggio JE (1993) Aluminum, iron, and zinc ions promote aggregation of physiological concentration of β-amyloid peptide. J. Neurochem. 61:1171-1174

Mills PJ, Ziegler MG, and Morrison TA (1998) Leptin is related to epinephrine levels but not reproductive hormone levels in cycling African-American and Caucasian women. Life Sci. 63:617-623

Pelleymounter MA, Cullen MJ, Healy D, Hecht R, Winters D, and McCaleb M (1998) Efficacy of exogenous recombinant murine leptin in lean and obese 10-to 12-mo-old female CD-1 mice. Am. J. Physiol. 275:R950-R959

Reidy SP and Weber JM (2002) Accelerated substrate cycling: a new energy-wasting role for leptin in vivo. Am. J. Physiology 282:E312-E317

Steinberg GR, Bonen A, and Dyck DJ (2002) Fatty acid oxidation and triacylglycerol hydrolysis are enhanced after chronic leptin treatment in rats. Am. J. Physiol. 282:E593-E600

Takekoshi K, Motooka M, Isobe K, Nomura F, Manmoku T, Ishii K, and Nakai T (1999) Leptin directly stimulates catecholamine secretion and synthesis in cultured porcine adrenal medullary chromaffin cells. Biochem. Biophys. Res. Commun. 261:426-431

Zhang Y, Proenca R, Maffel M, Barone M, Leopold L, and Friedman JM (1994) Positional cloning of the mouse obese gene and its human homologue. Nature 372:425-432

Ultrastructural Study on the Demyelination of the White Matter in the Rat Spinal Cord after Cardiac Arrest and Massive Hemorrhagic Shock

Li Yin[1], Yoshifumi Kudo[1, 2], Hirokazu Ohtaki[1], Sachiko Yofu[1, 3], Masaji Matsunaga[1, 3], and Seiji Shioda[1]

[1]Department of Anatomy and [2]Department of Orthopedic Surgery, Showa University School of Medicine, 1-5-8 Hatanodai, Shinagawa-Ku, Tokyo 142-8555, Japan
[3]Gene Trophology Research Institute, Tokyo 130-0012, Japan

Summary. Neuronal damages in lumbar spinal cord are known to induce the motor deficit after systemic ischemia including cardiac arrest and heart attack. Recently we reported that the neuronal damages in the hippocampus and lumbar spinal cord were induced after the combination of cardiac arrest and massive hemorrhagic shock. In the model, the neurons in intermediate gray matter were showed necrotic neuronal cell death at 1 d after ischemia. However, the damages of white matter did not understood well. Therefore, we focused on the white matter damages and have checked on the processes of the demyelination in limber spinal cord after severe systemic ischemia using electron microscopy. At 1 d, the myelin sheaths was collapsed and dissociated from axon although the axons seemed relatively intact. At 7 d, many shrank axons were observed and some of axons were disappeared. This result suggested the progressively demyelination was observed in ventral white matter in lumber spinal cord prior to axonal degeneration after severe systemic ischemia.

Key words. Global ischemia, spinal cord injury, demyelination, electron microscopy, axons

1 Introduction

Neuronal damages in lumbar spinal cord are known to induce the motor deficit after systemic ischemia including cardiac arrest and heart attack. The motor deficit as typified by hind limb paralysis distress many survivors even if they were well out of death after cardiac arrest (Imaizumi et al 1994). The clinical retrospective study suggests that lumbosacral spinal cord was the most vulnerable area to global ischemia (Duggal and Lach 2002). Recently we reported that the neuronal damages in the hippocampus and lumbar spinal cord were induced after severe systemic ischemia which is induced by cardiac arrest and massive hemorrhagic shock (Kudo et al 2005). In the model, the neurons in intermediate gray matter were showed necrotic neuronal cell death at 1 d after ischemia. However, the motor neuron in the anterior horn did not show the neuronal cell death in spite of the animals shows severe hind limb paralysis. On the other hands, it is reported the white matter was observed demyelination in the brain and spinal cord and leading to neurological paralysis in multiple sclerosis (Kanwar et al 2004; Bambakidis and Miller 2004). Therefore, we focused on the white matter damages and have checked on the processes of the demyelination in limber spinal cord after severe systemic ischemia using electron microscopy.

2 Demyelination in Lumber Spinal Cord After Global Ischemia

Rats were subjected to 7 min hypotension and 5 min cardiac arrest as a severe systemic ischemic model as previously described (Kudo et al 2005). At 0, 1 and 7 d after ischemia, we observed in ventral white matter in lumber spinal cord with electron microscopy. The myelinated axons at 0 d after ischemia were round and low electron density, and the myelin sheaths are smooth (*arrow* in Fig.1D), stratified regular shape, contacted with axon and wrapped around it. At 1 d, the myelin sheaths showed wide and irregular extracellular spaces, dissociated from axon (*arrows* in Fig. 1B, E) and noted in myelin ovoids (*arrowheads* in Fig.1B, E). However, the axons seemed relatively intact while the density in the cytoplasm slightly increased. These features were considered demyelination. At 7 d after ischemia, extensive demyelination was observed and the myelin sheaths showed fully irregular shape (*arrows* in Fig. 1C, F). Many shrank axons were observed and some of axons were disappeared.

3 Conclusions

This result suggested the progressively demyelination was observed in ventral white matter in lumber spinal cord prior to axonal degeneration after severe systemic ischemia. The demyelination in white matter of lumber spinal cord might be involving for hind limb paralysis after global ischemia.

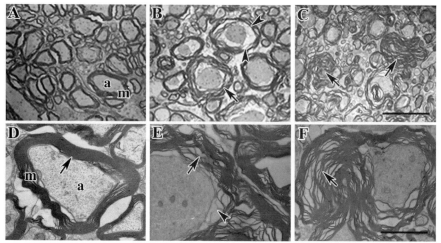

Fig. 1. Electron micrographs showing the ultrastructural changes in the ventral white matter of the lumbar spinal cord after ischemia. (**A, D**) Normal axonal (*a*) and myelin sheaths (*m*) structure. (**B, E**) Disintegrating myelin sheaths at 1d after ischemia. (**C, F**) Extensive demyelination at 7 d after ischemia. (*Scale bars*: 0.4μm **A-C**, 2μm **D-F**)

References

Bambakidis NC, Miller RH (2004) Transplantation of oligodendrocyte precursors and sonic hedgehog results in improved function and white matter sparing in the spinal cords of adult rats after contusion. Spine J 4:16-26

Duggal N, Lach B (2002) Selective vulnerability of the lumbosacral spinal cord after cardiac arrest and hypotension. Stroke 33:116-121

Imaizumi H, Ujike Y, Asai Y (1994) Spinal cord ischemia after cardiac arrest. J Emerg Med 12:789-793

Kudo Y, Ohtaki H, Dohi K, Yin L, Nakamachi T, Endo S, Yofu S, Miyaoka H and Shiod S (2005) Neuronal damage in rat brain and spinal cord after cardiac arrest with massive hemorrhagic shock. Crit Care Med (*in press*)

Kanwar JR, Kanwar RK, Krissansen GW (2004) Simultaneous neuroprotection and blockade of inflammation reverses autoimmune encephalomyelitis. Brain 127:1313-1331

Engagement of Proinflammatory Cytokines after Cerebral Ischemia

Hirokazu Ohtaki[1], Li Yin[1], Kenji Dohi[1, 2], Sachiko Yofu[1, 3], Tomoya Nakamachi[1], Masaji Matsunaga[3], and Seiji Shioda[1]

[1] Department of Anatomy, and [2] Emergency and Clinical Care Medicine Showa University School of Medicine, 1-5-8 Hatanodai, Shinagawa-ku Tokyo 142-8555, Japan
[3] Gene Trophology Research Institute, Tokyo 130-0012, Japan

Summary. The incidence of stroke is gradually increasing in the industrialized world and a major cause of long-lasting disability. To determine the mechanism of neuronal cell death after stroke and develop the new strategy of therapy, several studies had been performed during the past decade and some proinflammatory cytokines were identified as the therapeutic candidate. However, the critical actions of proinflammatory cytokines are still largely unknown in detail because the cytokines play a role on small amount, show the short half-life, and have multiple functions for depending on the amount and time of expression. We present here the pathophysiological events of cerebral ischemia and the action and possibility of proinflammatory cytokines from the recent results.

Key words. Ischemic, neuronal cell death, interleukin-1, tumor necrosis factor α, interleukin-6, oxidative stress

1 Pathophysiological Events of Cerebral Ischemia

Cerebrovascular disease is the third leading cause of death in the industrialized world and a major cause of long-lasting disability. The incidence of brain infarction (stroke) is gradually increasing, due to relatively recent life-style changes, such as consumption fatty foods, smoking, and excess stress.

Neuronal function and survivorship require appropriate supply of blood to the nervous system. The reduction of cerebral blood flow (CBF) below one fourth of the normal values, that is ischemia, is caused by the occlusion of a cerebral artery by an embolus and local thrombosis or by insufficient cir-

41

culation during cardiac arrest, and is disappeared immediately the neuronal activity. Brain injury after ischemia develops from a complex series of pathophysiological events that evolve with time and affect specific brain regions (Fig. 1). Primary neuronal cell death appears rapidly in the core region and is accompanied by the secondary death in the ischemic penumbra that evolves subsequent to activation of multiple death pathways.

The impairment of CBF restricts immediately the delivery of energy, O_2 and glucose, and, decreases the production of ATP. It results in a depolarization of the plasma membrane, a rapid loss of K^+ from neurons, massive influx of water with the influx of Na^+ and Cl^- and induction of cytotoxic edema. In addition, the brain ischemia causes a disruption of the blood brain barrier (BBB) and increases percolation of fluid from vessels into the brain parenchyma and results in vasogenic edema. This period also evokes excitotoxicity by mediating the glutamate receptors, and increases intracellular Ca^{2+} ($[Ca^{2+}]i$) which also accelerates the neuronal cell death (Choi 1988).

After a few hours to a few days, excess $[Ca^{2+}]i$ and glial cell activation causes the expression of the inflammatory cytokines, such as TNFα and IL-1β (Chen and Swanson 2003, Rothwell and Luheshi 2000, Shohami et al. 1999). These then play a central role in this subacute phase. The cytokines indirectly lead to the generation of reactive oxygen species (ROS) and the formation of excessive amounts of free radicals, including NO, super oxide anion (O_2^-), hydrogen peroxide (H_2O_2) and peroxynitrite ($ONOO^-$) (Chan 2001, Iadecola 1997, Mizushima et al. 2002, Ohtaki et al. 2003a,b). Elevated $[Ca^{2+}]i$ also increases uncoupled oxidative phosphorylation in the mitochondria, which also leads to a further decrease in energy supply and increase in ROS (Murakami et al. 1998) and activates series of degradative enzymes, such as lipases, proteases, endonucleases and other catabolic enzymes. The degradative enzymes collectively have detrimental consequences upon cell functions, membrane structure and the cytoskeleton. Even if blood flow is restored by reperfusion, the addition of oxygen can actually enhance the biochemical reactions present that generate ROS. In the ischemic zone, endothelial adhesion molecules are upregulated and leukocytes migrate through the walls of blood vessels, invade the parenchyma and release their inflammatory cytokines, NOS, and ROS (Iadecola 1997, Mizushima et al. 2002, Ohtaki et al. 2003a,b, 2004, Rothwell and Luheshi 2000, Yin et al. 2004).

After a few days (chronic phase), ischemic penumbra in the cells is ensnared in delayed neuronal cell death (DNCD) by TNFα and the oxidation of mitochondria. The DNCD includes in part the product of apoptotic processes such as the activation of caspases cascade (Dohi et al. 2003,

Namura et al. 1998). In conclusion, ischemic brain damage is multi-dimensional in origin and offers a broad range of targets for neuroprotective intervention.

Although the blockade of factors in acute phase should be promptly delivered to ischemic patients for appropriate efficacy, most patients are hospitalized after 3 hours form onset of symptoms. Hence, the medicines, which are able to extend therapeutic time window, is needed as the ischemic therapeutic strategy. In the point of view, the factors in subacute and early chronic period such as inflammatory cytokins, ROS and its related factors have been focusing as therapeutic targets.

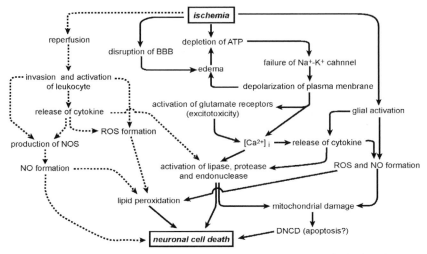

Fig. 1. Simplified overview of the neurotoxic pathway in brain ischemia. A complex neurotoxic pathway is caused by ischemia (*solid lines*). Some key events are the depolarization of the plasma membrane, over-activation of glutamate receptors and [Ca2+]i, the generation of cytokines and ROS, and the induction of apoptotic neuronal cell death. Even if the brain is reperfused (*dotted lines*), the cytokines and excess ROS induce neuronal cell death (Ohtaki et al. 2005)

2 IL-1

IL-1 consists of two molecular species, IL-1α and IL-1β, which are derived from two distinct genes located 50 kb apart on chromosome 2 of the mouse genome (D'Eutachio et al. 1987).

Injection of IL-1β (icv) is markedly exacerbated stroke size and injection of IL-1 receptor antagonist is improved it after ischemia (Allan and Rothwell 2001). IL-1β is increased in microglia/macrophage several hours after ischemia. Therefore, it is suggested that IL-1β plays mainly the exacerbation of neuronal injury. However, it has been revealed recently that IL-1α also plays a critical role for the induction of neuronal cell death because the IL-1 α/β KO mouse have increased significantly the infarct volume to compare with the wild-type mouse after focal ischemia, but not IL-1 α and β sole KO mouse (Boutin et al. 2002, Ohtaki et al. 2003a). There are two IL-1 receptor types, IL-1RI and RII, IL-1 believes to exert all of its actions by binding to IL-1RI (Dinarello 1996). As shown in Figs. 2 and 3 (see color page), the action of IL-1 is increased the ROS such as NO, ONOO⁻, O_2^-, and H_2O_2 by regulating the NO synthases (NOSs) and cyclooxygenase (Mizushima et al. 2002, Ohtaki et al. 2003a,b, Serou et al. 1999, Samad et al. 2001).

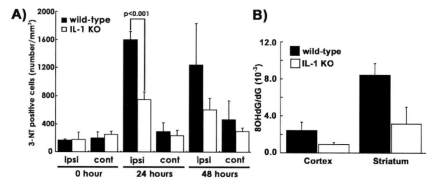

Fig. 2. Suppression of oxidative damages in IL-1 KO mouse after ischemia. A) is showing number of 3-nitro-L-tyrosine (3-NT), which is oxidative metabolite of ONOO-, immunopositive (+) cells. There is few 3-NT (+) cells in either post-ischemia (0 h) or in the contralateral hemisphere (cont). The 3-NT (+) cells in the ipsilateral hemisphere (ipsi) increase with time, peaking at 24 h then decreasing by 48 h after ischemia. The number of 3-NT (+) cells in the ipsilateral hemisphere of the wild-type mice is significantly higher than that measured in IL-1 KO mice at 24 h. B) is showing 8OHdG / dG ratio of the cortex and striatum at 24 h after ischemia. 8OHdG is marker of oxidative DNA damage by O_2^-, and H_2O_2. Brain 8OHdG/dG ratio of wild-type mice is higher than that of IL-1 KO mice at 24 h after ischemia. n=3-4 mice in each group

3 TNF-α

TNFα is known to participate in several central nervous system (CNS) disorders including Alzheimer's disease, multiple sclerosis, Parkinson's disease, and cerebral ischemia (Hallenbeck 2002). In neurodegenerative model, TNFα is suggested to relate in the demyelination and remyelination (Arnett et al. 2001). According to some studies by animal ischemia, TNFα is expressed biphasically during ischemia and after reperfusion in neurons, astroglia, microglia, oligodendrocytes, and endothelial cells in CNS tissue (Ohtaki et al. 2004, Botchkina et al. 1997, Gong et al. 1998, Lambertsen et al. 2002, Sairanen et al. 2001, Tseng et al. 1999, Uno et al, 1997) and is considered to deteriorate the neuronal cell death. There are two receptors, p55 (TNFRI) and p75 (TNFRII). TNFRI also increases after transient ischemia. The immunopositive cells of TNFRI are observed in the neuron and astrocytes on depending of time after ischemia (Botchkina et al. 1997, Yin et al. 2004). TNFRI is known to have at least three signal transductions which are caspases, IKK kinase and SAPK/JNK cascades. The caspases and SAPK/JNK cascades are participated in the induction of neuronal cell death. However, IKK kinase cascade is considered to play an important role for neuroprotection by generation of antioxidant such as superoxide dismutase (Shohami et al. 1999, Hallenbeck 2002). Recently, the infarct volume in TNFRI KO mice is larger than that in the wild-type mice (Bruce et al. 1996). Therefore, a critical role of TNFα and the signaling is controversial yet.

4 IL-6

Cerebral ventricular injection (icv) of IL-6 was reduced ischemic neuronal cell death (Loddick et al. 1998). However, IL-6 KO mice did not show the increases of the infarction after ischemia (Clark et al. 2000). Therefore, the critical role of IL-6 during ischemia has not understood yet. However, we recently determined evidence that IL-6 might be participated in the neuroprotective effect of a neuropeptide, pituitary adenylate cyclase-activating polypeptide (PACAP) (Arimura 1998, Arimura and Shioda 1995). The injection of PACAP is known to suppress the neuronal cell death after ischemia (Uchida et al. 1996, Reglodi et al. 2000). It is reported that PACAP stimulated IL-6 production in pituitary via cAMP accumulation (Tatsuno I et al. 1991). We have determined that icv injection of PACAP increased IL-6 in CSF after global ischemia in rat (Shioda et al. 1998). The result is suggesting that the neuroprotective effect of PACAP depends partially on IL-6.

5 Conclusion

It is doubtless for inflammatory cytokines to participate in the process of the neuronal cell death after ischemia. However, the mechanism of these actions are still largely unknown in detail because the cytokines play a role on small amount, show the short half-life, and have multiple functions for depending on the amount and time of expression. Some cytokines, the inhibitors, and modulators might be able to provide therapeutic targets by determining the action of the cytokines.

6 Acknowledgements

This study was supported in part by grants from the Ministry of Education, Science, Sports and Culture (H.O. and S.S.), and a High-Technology Research Center Project from the Ministry of Education, Science, Sports and Culture of Japan (S.S.).

References

Allan SM, Rothwell NJ (2001) Cytokines and acute neurodegeneration. Nat Rev Neurosci 2:734-744

Arimura A (1998) Perspectives on pituitary adenylate cyclase activating polypeptide (PACAP) in the neuroendocrine, endocrine, and nervous systems. Jpn J Physiol 48:301-331

Arimura A, Shioda S (1995) Pituitary adenylate cyclase activating polypeptide (PACAP) and its receptors: neuroendocrine and endocrine interaction. Front Neuroendocrinol 16:53-88

Arnett HA, Mason J, Marino M, Suzuki K, Matsushima GK, Ting JP (2001) TNFα promotes proliferation of oligodendrocyte progenitors and remyelination. Nat Neurosci 4:1116-1122

Botchkina GI, Meistrell ME 3rd, Botchkina IL, Tracey KJ (1997) Expression of TNF and TNF receptors (p55 and p75) in the rat brain after focal cerebral ischemia. Mol Med 3:765-781

Boutin H, LeFeuvre RA, Horai R, Asano M, Iwakura Y, Rothwell NJ. (2001) Role of IL-1α and IL-1β in ischemic brain damage. J. Neurosci 21:5528-5534.

Bruce AJ, Boling W, Kindy MS, Peschon J, Kraemer PJ, Carpenter MK, Holtsberg FW, Mattson MP (1996) Altered neuronal and microglial responses to excitotoxic and ischemic brain injury in mice lacking TNF receptors. Nat Med 2:788-794

Chan PH (2001) Reactive oxygen radicals in signaling and damage in the ischemic brain. J Cereb Blood Flow Metab 21:2-14

Chen Y, Swanson RA (2003) Astrocytes and brain injury. J Cereb Blood Flow Metab 23:137-149

Choi DW (1988) Calcium-mediated neurotoxicity: relationship to specific channel types and role in ischemic damage. Trends Neurosci 11:465-469.

Clark WM, Rinker LG, Lessov NS, Hazel K, Hill JK, Stenzel-Poore M, Eckenstein F. (2000) Lack of interleukin-6 expression is not protective against focal central nervous system ischemia. Stroke 31:1715-1720

D'Eutachio P, Jadidi S, Fuhlbrigge RC, Gray PW, Chaplin DD (1987) Interleukin-1α and β genes: linkage on chromosome 2 in the mouse. Immunogenetics 26:339-343

Dinarello CA (1996) Biologic basis for interleukin-1 in disease. Blood 87:2095-2147

Dohi K, Ohtaki H, Inn R, Ikeda Y, Shioda S, Aruga T (2003) Peroxynitrite and caspase-3 expression after ischemia/reperfusion in mouse cardiac arrest model. Acta Neurochir (Suppl.) 86:87-91

Gong C, Qin Z, Betz AL, Liu XH, Yang GY (1998) Cellular localization of tumor necrosis factor α following focal cerebral ischemia in mice. Brain Res 801:1-8

Hallenbeck JM (2002) The many faces of tumor necrosis factor in stroke. Nat Med 8:1363-1368

Iadecola C (1997) Bright and dark sides of nitric oxide in ischemic brain injury. Trends Neurosci 20:132-139

Lambertsen KL, Gregersen R, Finsen B (2002) Microglial-macrophage synthesis of tumor necrosis factor after focal cerebral ischemia in mice is strain dependent. J Cereb Blood Flow Metab 22:785-797

Loddick SA, Turnbull AV, Rothwell NJ (1998) Cerebral interleukin-6 is neuroprotective during permanent focal cerebral ischemia in the rat. J Cereb Blood Flow Metab 18:176-179

Mizushima H, Zhou CJ, Dohi K, Horai R, Asano M, Iwakura Y, Hirabayashi T, Arata S, Nakajo S, Takaki A, Ohtaki H, Shioda S (2002) Reduced postischemic apoptosis in the hippocampus of mice deficient in interleukin-1. J Comp Neurol 448:203-216

Murakami K, Kondo T, Kawase M, Li Y, Sato S, Chen SF, Chan PH (1998) Mitochondrial susceptibility to oxidative stress exacerbates cerebral infarction that follows permanent focal cerebral ischemia in mutant mice with manganese superoxide dismutase deficiency. J Neurosci 18:205-213

Namura S, Zhu J, Fink K, Endres M, Srinivasan A, Tomaselli KJ, Yuan J, Moskowitz MA (1998) Activation and cleavage of caspase-3 in apoptosis induced by experimental cerebral ischemia. J Neurosci 18:3659-3668

Ohtaki H, Dohi K, Nakamachi T, Yofu S, Endo S, Kudo Y, Shioda S (2005) Evaluation of Brain Ischemia in Mice. Acta Histochem Cytochem 38:99-106

Ohtaki H, Funahashi H, Dohi K, Oguro T, Horai R, Asano M, Iwakura Y, Yin L, Matsunaga M, Goto N, Shioda S (2003a) Suppression of oxidative neuronal damage after transient middle cerebral artery occlusion in mice lacking interleukin-1. Neurosci Res. 45:313-324

Ohtaki H, Takaki A, Yin L, Dohi K, Nakamachi T, Matsunaga M, Horai R, Asano M, Iwakura Y, Shioda S. (2003b) Suppression of oxidative stress after transient focal ischemia in interleukin-1 knock out mice. Acta Neurochir (Suppl.) 86:191-194

Ohtaki H, Yin L, Nakamachi T, Dohi K, Kudo Y, Makino R, Shioda S (2004) Expression of tumor necrosis factor α in nerve fibers and oligodendrocytes after transient focal ischemia in mice. Neurosci Lett 368:162-166

Reglodi D, Somogyvari-Vigh A, Vigh S, Kozicz T, Arimura A (2000) Delayed systemic administration of PACAP38 is neuroprotective in transient middle cerebral artery occlusion in the rat. Stroke 31:1411-1417

Rothwell NJ, Luheshi GN (2000) Interleukin 1 in the brain: biology, pathology and therapeutic target. Trends Neurosci. 23:618-625

Sairanen TR, Lindsberg PJ, Brenner M, Carpen O, Siren A (2001) Differential cellular expression of tumor necrosis factor α and type I tumor necrosis factor receptor after transient global forebrain ischemia. J Neurol Sci 186:87-99

Samad TA, Moore KA, Sapirstein A, Billet S, Allchorne A, Poole S, Bonventre JV, Woolf CJ (2001) Interleukin-1β-mediated induction of COX-2 in the CNS contributes to inflammatory pain hypersensitivity. Nature 410:471-475

Serou MJ, DeCoster MA, Bazan NG (1999) Interleukin-1β activates expression of cyclooxygenase-2 and inducible nitric oxide synthase in primary hippocampal neuronal culture: platelet-activating factor as a preferential mediator of cyclooxygenase-2 expression. J Neurosci Res 58:593-598.

Shioda S, Ozawa H, Dohi K, Mizushima H, Matsumoto K, Nakajo S, Takaki A, Zhou CJ, Nakai Y, Arimura A (1998) PACAP protects hippocampal neurons against apoptosis: involvement of JNK/SAPK signaling pathway. Ann N Y Acad Sci 865:111-117

Shohami E, Ginis I, Hallenbeck JM (1999) Dual role of tumor necrosis factor α in brain injury. Cytokine Growth Factor Rev 10:119-130

Tatsuno I, Somogyvari-Vigh A, Mizuno K, Gottschall PE, Hidaka H, Arimura A. (1991) Neuropeptide regulation of interleukin-6 production from the pituitary: stimulation by pituitary adenylate cyclase activating polypeptide and calcitonin gene-related peptide. Endocrinology 129:1797-804

Tseng MT, Chang CC (1999) Ultrastructural localization of hippocampal TNF-α immunoreactive cells in rats following transient global ischemia. Brain Res 833:121-124

Uchida D, Arimura A, Somogyvari-Vigh A, Shioda S, Banks WA (1996) Prevention of ischemia-induced death of hippocampal neurons by pituitary adenylate cyclase activating polypeptide. Brain Res 736:280-286

Uno H, Matsuyama T, Akita H, Nishimura H, Sugita M (1997) Induction of tumor necrosis factor-α in the mouse hippocampus following transient forebrain ischemia. J Cereb Blood Flow Metab 17:491-499

Yin L, Ohtaki H, Nakamachi T, Kudo Y, Makino R, Shioda S (2004) Delayed expressed TNFR1 co-localize with ICAM-1 in astrocyte in mice brain after transient focal ischemia. Neurosci Lett 370:30-35

Levels of the Alkoxy Radical in Patients with Brain Death

Kenji Dohi[1], Kazue Satoh[2], Hiroshi Moriwaki[1], Yuko Mihara[1],
Yasufumi Miyake[1], Hirokazu Ohtaki[2], Seiji Shioda[2], and Tohru Aruga[1]

[1]Department of Critical Care and Emergency Medicine, [2]Department of
Anatomy, School of Medicine, Showa University, 1-5-8 Hatanodai, Shinagawa-ku, Tokyo 142-8555, Japan

Summary. Lipid peroxidation is induced by reactive oxygen species and is
involved in acute neuronal damage. Thus, controlling this process may be
a realistic therapeutic strategy for treating or preventing neuronal damage.
However, the study of free radicals in body fluids is currently severely
hampered by technical difficulties in their detection. We have developed an
ex vivo electron spin resonance (ESR) method, that employs 5,5-dimethyl-
1-pyrroline-N-oxide (DMPO) as a spin trap, to measure the alkoxyl radical
(OR˙) in human blood. We found that this method can detect OR˙ produced
by treating human blood *ex vivo*. Analysis of jugular venous and radial arterial blood from patients with acute neuronal damage revealed higher OR˙
levels in the venous blood compared to arterial samples. In brain death patients, however, the jugular vein OR˙ spectrum was similar to the radial artery OR˙ level. The ratio between radial artery and jugular vein levels, expressed as the RI (radical intensity) ratio = aORI/jORI, was calculated.
Arterial-jugular bulb RI ratio below 1, together with accepted clinical criteria (unresponsive coma with brainstem areflexia), may provide a non-
invasive assessment of cerebral circulatory arrest that can help to predict
brain death. This novel *ex vivo* ESR method may be very useful for measuring oxidative stress in patients with acute neuronal damage.

Key words. Free radical, free radical scavenger, reactive oxygen species,
ROS, electron spin resonance, alkoxy radical, brain ischemia, traumatic
brain injury, human, monitoring, edaravone

1 Introduction

Oxidative stress plays a key role in both primary and secondary damage following acute neuronal injury (Ikeda et al. 1990) (Juurlink et al. 1998) (Tyurin et al. 2000). Reactive oxygen species (ROS) have been shown to play important roles in various neuronal conditions (Lewen et al. 2000) (Lewen et al. 2001a). Oxidative stress causes lipid peroxidation, via activation of reactive oxygen species, and this process is associated with brain damage and acute neuronal injury (Watanabe et al. 1994a) (Watanabe et al. 1994b) (Tyurin et al. 2000). Thus, controlling lipid peroxidation is an important strategy in the treatment and/or prevention of neuronal damage. In animal experiments, free radical scavengers and antioxidants have been shown to dramatically reduce cerebral damage (Lewen et al. 2001b) (Nakamura et al. 2003) (Watanabe et al. 1994b). However, clear clinical corroboration of this animal evidence has been rather slow to emerge, and those results obtained to date have been somewhat ambiguous (Houkin et al. 2003). The alkoxy radical (OR˙) is an important chemical and biological entity, being involved in many radical chain reactions, including lipid peroxidation (Blair 2001). We previously developed a novel method for analyzing OR˙ by *ex vivo* ESR (Mihara et al 2004).

The diagnosis of brain death is generally based on both clinical and laboratory findings (Aruga 2000). In Japan, brain death must also include death of the brain stem, and its diagnosis is based on both clinical and electrophysiological criteria. However, the diagnosis of brain death is still contentious and reliable biomarkers and tests are required.

In this present study, we analyzed OR˙ levels in blood from the jugular vein and radial artery in patients with brain death, using our recently developed *ex vivo* ESR method.

2 Patients and Methods

2.1 Materials

5, 5-dimethyl-l-pyroline-N-oxide (DMPO) was from Dojin Ltd., Kumamoto, Japan and diethylenetriamine penta-acetic acid (DETAPAC) was from Wako Pure Chem Ind., Ltd., Osaka.

2.2 Patient Population

Patients with brain death (n = 9) admitted to Showa University Hospital Emergency Centre were enrolled in the study. The primary diseases causing brain death in the study population were severe neurotrauma, cardiopulmonary arrest and stroke.

Evaluation of brain death conformed to Japanese guidelines and involved routine examination of various clinical parameters including brain stem reflex, pupil dilatation, electroencephalogram (EEG) and auditory brainstem response (ABR) (Aruga 2000).

2.3 Blood Sampling

Blood samples were collected from catheters in the bulb of the jugular vein and in the radial artery before and after brain death diagnosis.

2.4 Assay for OR·

After collection, whole blood samples were immediately mixed with a spin trap, 5, 5-dimethyl-1-pyroline-N-oxide (DMPO), which was purchased from Dojin Chemical (Tokyo, Japan), and analyzed within one hour. The ESR spectra were recorded at room temperature （23℃） with a spectrometer (JEOL JES REIX, X-band, 100kHz modulation frequency). Instrument settings were as follows: centre field, 335.0 ± 5.0mT; microwave power, 5mW; modulation amplitude, 0.1mT; gain, 200; time constant, 0.1s; scanning time, 2min.

Radical intensity (RI) was defined as the ratio of signal intensity of the first peak of the alkoxyl-DMPO spin adduct to that of MnO (external standard).

2.5 Comparison of Blood OR· Levels in Jugular Vein and Radial Artery

We expressed the difference between OR· levels in the jugular vein and radial artery as the RI (ORIj/ORIa) ratio. RI ratio values before and after diagnosis of brain death were compared by Student's t test. Differences between means were considered statistically significant when the P value was less than 0.05.

3 Results

3.1 OR˙ Levels in Jugular Vein and Radial Artery Blood

Alkoxy radical spin adducts were observed in all samples (Fig. 1a). The radical adduct was chemically identified as the alkoxy radical by adding dimethyl sulfoxide or catalase in the ESR experiments (data not shown). Composite computer simulations of the alkoxy radical adduct are shown in Fig.1b.

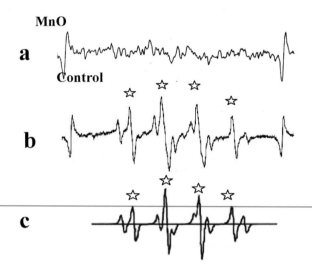

Fig. 1. Radical adduct from a healthy volunteer and a neurotrauma patient (**a, b**). **a**: Radical adduct from a healthy volunteer. **b**: radical adduct from a severe neurotrauma patient. **c**: Computer simulation of radical adducts. *Star*: spin adduct of the alkoxy radical

We demonstrated alteration of the RI ratio in one typical patient, who had been diagnosed with acute subdural hematoma. Before the diagnosis of brain death (on day 3), the OR˙ level in jugular vein blood was much higher than radial artery blood (Fig. 2a,b), with an RI ratio of 2.0. After brain death diagnosis (on day 6), the OR˙ levels in the jugular vein and radial artery were almost equivalent, such that the RI ratio decreased to 1.0 (Fig. 2c,d). Over the 9 brain death cases, the average RI ratio was 0.92 (Fig. 3). Although the mean OR level was lower in the jugular vein than the radial artery, this difference was not significant by Student's t-test (p = 0.342).

4 Discussion

In the present study, we reported whole blood OR levels in patients with brain death. Notably, prior to brain death, the jORI was higher than the aORI, but a decrease in RI ratio was observed after diagnosis of brain death.

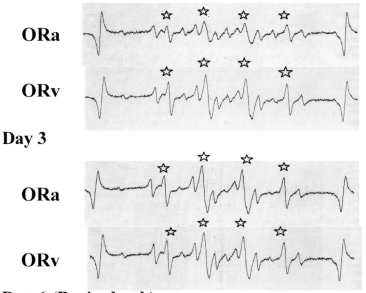

Fig. 2. Radical adduct in jugular venous blood before (*Day3*) and after diagnosis (*Day 6*) of brain death in the patient with neurotrauma

On day3, jORI was much higher than that of aORI (upper). On the status of brain death, ORI and radical adduct in juglar vein were almost similar with in radial artery (lower).

This diagnosis was conducted in agreement with Japanese legal requirements and involved assessment of consciousness level, brainstem reflexes, EEG, ABR and apnea.

Recently, it was reported that an increase in venous oxygen saturation in the jugular bulb (SjO2) is a useful marker for the diagnosis of brain death (Diaz-Reganon G et al 2002). A central venous-jugular bulb oxygen saturation rate below 1, together with accepted clinical criteria, provides non-invasive assessment of cerebral circulatory arrest that can help to predict brain death (Hayashi et al 2003). Elevation of SjO2 is thought to reflect decreased perfusion of cerebral blood flow and to involve a collateral vessel. Single photon emission computed tomography and cerebral angiography have been proposed as alternative methods for the detection of cerebral circulatory arrest (Hayashi et al 2003). It is possible that the results obtained in the current study might also reflect the stagnation of blood flow caused by elevation of intracranial pressure (ICP).

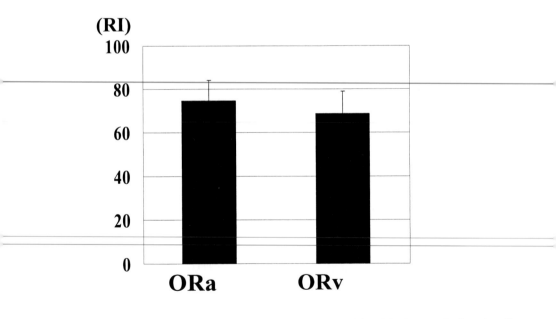

Fig. 3. Alkoxy radical intensities in jugular venous blood before and after the diagnosis of brain death (n = 9). BARS = SE. No statistical difference was observed

Ex vivo ESR method was possible to catch the slight changes of OR· immediately. Since the biological half-life of OR· is very short, it may be possible to compare jORI and aORI.

OR· is thought to be intimately associated with lipid peroxidation. It is formed *in vivo* via free radical pathways from the action of ROS on polyunsaturated acids. OR· is also a specific product of lipid hydroperoxides, lipoxygenases and cyclooxygenases (Blair 2001) (Pryor et al. 1990). These reactions mediate DNA damage and inflammation (Blair 2001). Lipid peroxidation is also known to be important in TBI (Tyurin et al. 2000).

Direct monitoring of free radicals would provide substantial pathophysiological information for the clinician. Recently, we developed a novel method for analysis and direct monitoring of OR· in human blood using an *ex vivo* ESR spin trap technique (Mihara et al. 2004). In that study, we reported that healthy volunteers had a detectable alkoxy radical intensity of 31.9 ± 19.5: (mean $\pm$ SD) (Mihara et al. 2004), which is significantly lower than the average OR· level observed in the current study (71.1 ± 6.5 : mean $\pm$ SE). It has also been reported that the OR· level is significantly lower in arterial blood than jugular venous blood in TBI patients, indicating that TBI augments the OR· level (Mihara et al. 2004). While the precise sources of OR· were not elucidated in that study, in the current report, we demonstrate that blood oxidation results in OR· production. Our results also support the contention that elevated levels of OR· in the jugular blood reflect the oxidative stress caused by damage to brain tissues and endovascular cells. Thus, the applications described here suggest that in addition to clarifying the pathophysiology of TBI, the ESR methodology we have developed may also be useful for monitoring oxidative stress.

In conclusion, this novel *ex vivo* ESR method is very useful for overcoming the technical difficulties associated with monitoring oxidative stress. Moreover, it may also be very valuable in predicting brain death in the clinical setting.

References

Aruga, T. (2000) The social problem of Brain death in Japan. The Japanese Journal of Acute Medicine 24: 1858-1862

Blair, I. A. (2001) Lipid hydroperoxide-mediated DNA damage. Exp Gerontol 36: 1473-1481.

Diaz-Reganon G, Minambres E, Holanda M., et al (2002) Usefulness of venous oxygen saturation in the jugular bulb for the diagnosis of brain death: report of 118 patients. Intensive Care Med 28: 1724-1728

Hayashi M, Toyoda I, Dohi K, et al (2003) Usefulness of cerebral angiography for the diaganosis of brain death. Brain Death and Cerebral Resuscitation 15: 40-45

Ikeda Y, Long DM, (1990) The molecular basis of brain injury and brain edema: the role of oxygen free radicals. Neurosurgery 27: 1-11

Juurlink B H, Paterson PG. (1998) Review of oxidative stress in brain and spinal cord injury: suggestions for pharmacological and nutritional management strategies. J Spinal Cord Med 21: 309-334

Lewen, A., Matz, P., and Chan, P. H., (2000) Free radical pathways in CNS injury. J Neurotrauma 17: 871-890

Lewen A, Fujimura M, Sugawara T, et al. (2001a) Oxidative stress-dependent release of mitochondrial cytochrome c after traumatic brain injury. J Cereb Blood Flow Metab 21: 914-920

Mihara Y, Dohi K, Satoh K, et al. (2004) Mild brain hypothermia suppresses oxygen free radical in patients with neuroemergency: An ex vivo electron spin resonance study. In: Hayashi N (ed) Hypothermia for acute brain damage. Springer-Verlag, Tokyo, pp 94-97

Ochi H, Morita I, Murota S (1992) Mechanism for endothelial cell injury induced by 15-hydroperoxyeicosatetraenoic acid, an arachidonate lipoxygenase product. Biochim Biophys Acta 1136: 247-252

Pryor WA, Porter NA (1990) Suggested mechanisms for the production of 4-hydroxy-2-nonenal from the autoxidation of polyunsaturated fatty acids. Free Radic Biol Med 8: 541-543

Tyurin VA, Tyurina YY, Borisenko GG, et al. (2000) Oxidative stress following traumatic brain injury in rats: quantitation of biomarkers and detection of free radical intermediates. *J Neurochem* 75:2178-2189

Yamato M, Egashira T, Utsumi H (2003) Application of in vivo ESR spectroscopy to measurement of cerebrovascular ROS generation in stroke. Free Radic Biol Med 35: 1619-1631

Differentiation of Neural Stem Cells into Astrocytes by Low Concentration of Pituitary Adenylate Cyclase-Activating Polypeptide (PACAP)

Jun Watanabe[1,2], Seiji Shioda[1], Sakae Kikuyama[2], Kazuyasu Nakaya[3], and Shigeo Nakajo[3]

[1]Department of First Anatomy, School of Medicine, Showa University, 1-5-8 Hatanodai, Shinagawa-ku, Tokyo 142-8555, Japan
[2]Department of Biology, School of Education, Waseda University, 1-6-1 Nishi-Waseda, Shinjuku-ku, Tokyo 169-8050, Japan
[3]Laboratory of Biological Chemistry, School of Pharmaceutical Sciences, Showa University, 1-5-8 Hatanodai, Shinagawa-ku, Tokyo 142-8555, Japan

Summary. Recently it has been reported that pituitary adenylate cyclase-activating polypeptide (PACAP) possesses diverse physiological functions in central nervous system. In this study, the effect of physiological concentrations of PACAP on the differentiation of neural stem cells is described. When neural stem cells were exposed to 0.2 nM PACAP, total number of the cells decreased during the 8 days-culture. However, number of MAP2-immunopositive cells decreased while GFAP-immunopositive cells increased depending on the concentration of PACAP (0.2-2 nM). Neural stem cells were differentiated into astrocytes even at a concentration below 2 nM. These results indicate that PACAP at a low concentration prefers induction of differentiation into astrocytes rather than proliferation in neural stem cells.

Key words. Neural stem cell, progenitor cell, differentiation, astrocyte, pituitary adenylate cyclase-activating polypeptide

1 Introduction

Neural progenitor cells possessing pluripotent and self-renewing properties undergo growth and differentiation in response to various stimuli (Gage et al. 1995, Kuhn and Svendsen 1999, Momma et al. 2000). It has previously been reported that neural stem cells can be differentiated into astrocytes by ciliary neurotrophic factor (CNTF), into oligodendrocytes by thyroid hormone T3, and into neurons by brain-derived neurotrophic factor (BDNF) and plate-derived growth factor (PDGF) (Johe et al. 1996). Basic fibroblast growth factor (bFGF) promotes self-proliferation of neural stem cells prepared from embryo.

PACAP has been purified from ovine hypothalamus (Miyata et al. 1989) and is known as a multifunctional neuropeptide (Arimura 1998, Vaudry et al. 2000). In recent studies, it has been shown that PACAP promotes differentiation of rat embryonic neural stem cells into astrocytes (Vallejo and Vallejo 2002, Ohno et al. 2005) and self-renewing in adult mouse neural stem cells (Mercer et al. 2004). In this study, the effect of PACAP at a lower concentration on mouse neural stem cells was investigated.

2 Materials and Methods

2.1 Materials

DMEM/F12, insulin, apotransferrin, and monoclonal antibodies for microtuble associated protein 2 (MAP2), galactocerebroside (Gal-C) and glial fibrillary acidic protein (GFAP) were obtained from Sigma Co (St. Louis, MO). Alexa Fluor^{TM} 546 goat anti-mouse IgG (H + L), PACAP, basic fibroblast growth factor (bFGF) and anti-nestin monoclonal antibody were purchased from Molecular Probes (Eugene, OR), Peptide Institute (Osaka), R & D systems (MN) and BD Biosciences Clontech (NJ), respectively.

2.2 Preparation and Culture of Neural Stem Cells

The cell culture was conducted according to the method described previously (Johe et al. 1996). Telencephalons were prepared from embryos (E14.5) of ICR mice and dissociated by gentle pipetting. The dissociated cells which contained neural progenitor cells were seeded on culture dishes previously coated with 15 µg/ml poly-L-ornithine and 1 µg/ml fibronectin. The cells were then cultured in the medium of DMEM/F12 containing 25 µg/ml insulin, 0.1 mg/ml apotransferrin, 20 nM progesterone, 0.1 mM putrescine, 30 nM sodium selenite, and 10 ng/ml of recombinant human bFGF for 4 days under 5% CO_2 and 95% air, followed by exposure (1-1.5

x 10^6 cells/10 cm dish) to various concentrations of PACAP.

2.3 Cell Identification

To identify the types of cells that differentiated from neural stem cells, immunofluorescent staining was carried out. Specific antibodies such as anti-MAP2 for neurons, anti-GFAP for astrocytes and anti-Gal C for oligodendrocytes were used as primary antibodies after the cells were exposed to various agents for the indicated times. The cells were fixed with 4% paraformaldehyde in phosphate–buffered saline (PBS) for 30 min, washed twice with PBS, and then blocking was performed with 3% BSA/PBS for 30 min. After cells had been washed twice with PBS, they were incubated with 0.1% TritonX-100/PBS for 30 min, then reacted overnight with primary antibodies (1:500 dilution in all cases except for 1:1000 dilution for anti-GFAP antibody) at 4°C after washing with PBS to remove the TritonX-100. Following the overnight incubation, cells were washed 4 times with PBS, and then incubated for 1 h at room temperature with 2 μg/ml Alexa 546 goat anti-mouse IgG (H + L). Dishes were mounted on the stage of a fluorescence microscope (Olympus BX50) and cell numbers were counted following staining of nuclei with Hoechst 33258.

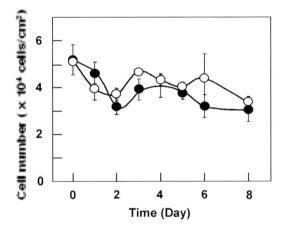

Fig. 1. Effect of PACAP on cell proliferation of neural stem cells. Neural stem cells were cultured with bFGF for 4 days, then bFGF were withdrawn from the culture medium and the cultures were restarted in the presence (*closed circle*) or absence (*open circle*) of 0.2 nM PACAP for the indicated times. Each vale indicates mean ± SD (n = 3)

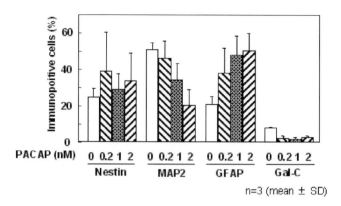

Fig. 2. Effect of low concentration PACAP on differentiation of neural stem cells. Mouse neural stem cells were treated with various concentration of PACAP for 8 days and then subjected to immunofluorescent staining. Each value indicates mean ± SD (n = 3)

3 Results

Neural stem cells were exposed to 0.2 nM PACAP withdrawing bFGF and cultured continually for up to 8 days (Fig. 1). There was almost no effect of PACAP on cell proliferation compared with that of control. Recently we have shown that PACAP at the concentration of 2 nM promotes not only induction of differentiation but also significant cell growth (Ohno et al. 2005).

Effect of lower concentrations of PACAP on differentiation of neural stem cells was investigated (Fig. 2). Number of MAP2-expressing cells decreased in dose dependent manner of PACAP. Inversely, GFAP-immunopositive cells increased with increasing concentration of PACAP. Moreover, the number of cells increased approximately 2-fold by treatment with 0.2 nM PACAP. This result indicates that neural stem cells possess the astrocyte-inducing ability even in concentrations lower than 2 nM. The morphological change of cells was observed by immunofluorescent staining (Fig. 3). Gal-C-immunopositive cells could be scarcely detected, whereas some cells reacted with other antibodies. As shown in Figure 3d, antibody against GFAP recognized typically fibrous component in cytoplasm of mouse neural stem cells.

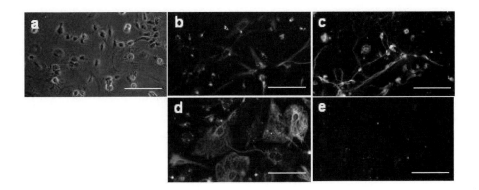

Fig. 3. Differentiation of neural stem cells into astrocytes by 0.2 nM PACAP. Neural stem cells were exposed to 0.2 nM PACAP for 8 days and observed under phase contrast microscope (**a**). Immunofluorescent staining was carried out using specific antibodies against nestin (**b**), MAP2 (**c**), GFAP (**d**) and Gal-C (**e**), respectively. *Scale bar* represents 50 μm

4 Discussion

In a recent study, we have found PACAP receptor, PAC1 on the plasma membranes of mouse neural stem cells by analyzing the expression levels of gene and protein. PACAP at the concentration of 2 nM bound to the PAC1 receptor with high affinity, and promoted not only differentiation of neural stem cells into astrocytes but also cell proliferation (Ohno et al. 2005).

In this experiment, the effect of PACAP at the lower concentrations on neural stem cells was investigated. The morphological changes were observed after around 4 days of exposure to PACAP. As shown in Figures 2 and 3, neural stem cells respond to even at low PACAP concentration (0.2 nM) and were differentiated into astrocytes. It is well known that there is three types of receptor, PAC1, VPAC1, and VPAC2 (Spengler et al. 1993, Chatterjee et al. 1996, Nicot and DiCicco-Bloom 2001). It is considered that this neuropeptide binds to PAC1 first, since PAC1 possesses the highest affinity for PACAP, then the signal is transmitted into cytoplasm. PAC1 is a receptor associated with G-protein. It is likely that the signal from PAC1 might be transmitted to Gq protein rather than Gs protein as described previously (Ohno et al. 2005).

Recently, it has been shown that nanomolar concentration of PACAP stimulates the proliferation of neural stem cells in adult mouse brain (Mercer et al. 2004) and embryonic neural stem cells (Ohno et al. 2005). However 0.2 nM PACAP did not influence the growth in embryonic neural stem cells as shown in Figure 1. Although there is no evidence to explain the precise molecular mechanism, it might be regulated by degree of stimulation of ligand itself.

It was demonstrated that low concentration (below 2 nM) PACAP has sufficient ability to induce differentiation to astrocytes in mouse embryonic neural stem cells. Further study will be performed to clarify the physiological significance in embryonic neural stem cells.

5 Acknowledgment

This work was supported in part by a Showa University Grant-in-Aid for Innovative Collaborative Research Projects.

References

Arimura A (1998) Perspectives on pituitary adenylate cyclase activating polypeptide (PACAP) in the neuroendocrine, endocrine, and nervous systems. Jpn J Physiol 48:301-331

Chatterjee TK, Sharma RV, Fisher RA (1996) Molecular cloning of a novel variant of the pituitary adenylate cyclase-activating polypeptides (PACAP) receptor that stimulates calcium influx by activation of L-type calcium channels. J Biol Chem 271:32226-32232

Johe KK, Hazel TG, Muller T, Dugich-Djordjevic MM, McKay RD (1996) Single factors direct the differentiation of stem cells from the fetal and adult central nervous system. Genes Dev 10:3129-3140

Mercer A, Rönnholm H, Holmberg J, Lundh H, Heidrich J, Zachrisson O, Ossoinak A, Frisén J, Patrone C (2004) PACAP promotes neural stem cell proliferation in adult mouse brain. J Neurosci Res 76:205-215

Miyata A, Arimura A, Dahl RR, Minamino N, Uehara A, Jiang L, Culler MD, Coy DH (1989) Isolation of a novel 38 residue-hypothalamic polypeptide which stimulates adenylate cyclase in pituitary cells. Biochem Biophys Res Commun 164:567-574

Nicot A, DiCicco-Bloom E (2001) Regulation of neuroblast mitosis is determined by PACAP receptor isoform expression. Proc Natl Acad Sci USA 98:4758-4763

Ohno F, Watanabe J, Sekihara H, Hirabayashi T, Arata S, Kikuyama S, Shioda S, Nakaya K, Nakajo S (2005) Pituitary adenylate cyclase-activating polypeptide (PACAP) promotes differentiation of mouse neural stem cells into astrocytes. Regul Peptides 126:115-122

Spengler D, Waeber C, Pantaloni C, Holsboer F, Bockaert J, Seeburg PH, Journot L (1993) Differential signal transduction by five splice variants of the PACAP receptor. Nature 365:170-175

Vallejo I, Vallejo M (2002) Pituitary adenylate cyclase-activating polypeptide induces astrocyte differentiation of precursor cells from developing cerebral cortex. Mol Cell Neurosci 21:671-683

Vaudry D, Gonzalez BJ, Basille M, Yon L, Fournier A, Vaudry H (2000) Pituitary adenylate cyclase-activating polypeptide and its receptors: from structure to functions. Pharmacol Rev 52:269-324

Neurogenesis is Toxicologically Heterogeneous: A Case of BrdU-Induced Cell Death

Tetsuo Ogawa, Makiko Kuwagata, and Seiji Shioda

Department of Anatomy, Showa University School of Medicine, 1-5-8 Hatanodai, Shinagawa-ku, Tokyo 142-8555, Japan

Summary. In developmental neuroscience, 5-bromo-2'-deoxyuridine (BrdU) is extensively used as a useful tool for labeling proliferating cells. Doses ranging from 10 to 100 mg/kg are commonly injected into animals. However, several lines of evidence demonstrated that BrdU has genotoxicity. The present study demonstrates that 1) a single injection of BrdU at 100 mg/kg into pregnant mice on gestation day 11.5 induced cell death in the neuroepithelium of certain brain areas such as the frontal neocortex, but not other areas such as the mesencephalic tegmentum and pons, 2) injections of BrdU into adult mice at 100 or 300 mg/kg did not induce cell death in the subventricular zone (SVZ) of the lateral ventricle, known to be a site of adult neurogenesis. These findings strongly suggest that 1) sensitivity to the genotoxicity differs among fetal neural stem cells, 2) BrdU-labeling should be avoided when early neurogenesis is being observed.

Key words. BrdU, neurogenesis, cell death, developmental neurotoxicology

1 Introduction

The thymidine analogue 5-bromo-2'-deoxyuridine (BrdU) is incorporated into the DNA as 5-bromouracil during the synthesis phase of the cell cycle of any cell, and has been used extensively as a useful tool for labeling proliferating cells in developmental neuroscience (Miller and Nowakowski, 1988; Soriano and Del Rio, 1991; Takahashi et al., 1999). However, several lines of evidence have indicated that BrdU has genotoxicity (reviewed by Morris, 1991). Mutagenesis occurs when BrdU triphosphate is incorporated into DNA, and when the BrdU-containing DNA replicates. Interestingly, prenatal exposure to BrdU, even at a dose generally administered to adult rodents (50 mg/kg), is reported to induce behavioral abnormalities such as locomotor hyperactivity (Kuwagata et al., 2004). In the present study, we demonstrate that BrdU induces cell death in the neuroepithelial layer of the several parts of the fetal brain, but not in the area of adult neurogenesis, and therefore neurogenesis is toxicologically heterogenous.

2 Methods

Pregnant animals were obtained by housing females with males (1-2 females /male). The time when a vaginal plug was observed in the morning was designated as gestation day 0.5 (GD0.5) and embryonic day 0.5 (E0.5). Pregnant mice on GD 11.5 were intraperitoneally administered 100 mg/kg of BrdU (Sigma, St. Louis, MO) or saline. Adult mice (9 week-old) were administered BrdU at a dose of 100 or 300 mg/kg.

Fetal brains (four fetuses from two litters) were prepared as described previously (Ogawa et al., 2005a,b). Briefly, the tissues were immersed in 4 % paraformaldehyde at 4°C for 2 days. The specimens were then embedded in 10 % gelatin and coronal sections were cut at a thickness of 45 μm in a vibratome. All of the serial sections were collected. Every third section was incubated in a solution of 10 mM PBS containing 5 % normal rabbit serum and 0.3 % Triton X-100 overnight, mounted on slide glass and processed for Nissl staining. Some of the other sections were subjected to BrdU immunohistochemistry or TUNEL staining. Like the fetal brains, adult brains (N = 6 / group) were also sectioned after perfusion fixation, and sections including the lateral ventricle were subjected to BrdU immunohistochemistry and TUNEL staining.

3 Results and Discussion

Exposure to BrdU at a dose of 100 mg/kg, which is generally administered in neurodevelopment studies, to E11.5 mice induced cell death in the neuroepithelial layers. Observation of a variety of brain areas on serial sections demonstrated variation in severity among the regions (Table 1). Furthermore, marked TUNEL-positive reactivity was observed in the BrdU-treated brain (Fig. 1), suggesting that the BrdU-induced cell death includes DNA strand breaks. By contrast, cell death was not induced by BrdU in the SVZ of the lateral ventricle in the adult brain despite evidence of neurogenesis detected by BrdU immunoreactivity (data not shown).

We have reported that abnormal cortical plate, as well as cell death was induced in the fetal rat brain in a BrdU-induced hyperactivity disorder

Table 1. Induction of cell death in the neuroepithelium of the fetal mouse brain after BrdU exposure

Neocortex	Septum	Striatum	Striatum (d)	Pallidum	Preoptic area
+++	−~±	+	−	+	±

Hippocampus	Thalamus	Amygdala	Pretectum	Mesephalic tegmentum
−~±	+	±	−~±	−

Pons	Superior colliculus	Inferior colliculus	Cerebellum
−	±	±	−

d, differentiating zone.
± slight; + mild; ++ moderate; +++ marked.

Model (Ogawa et al., 2005b). In the rat study, we observed serial sections containing a wide variety of fetal brain areas (over 20 areas) after BrdU-exposure on gestation days 9 to 15, and abnormalities were observed in the forebrain, but not in the brain stem (Ogawa et al., 2005b). In the present study, we used a single dosage and evaluated the effect of this compound

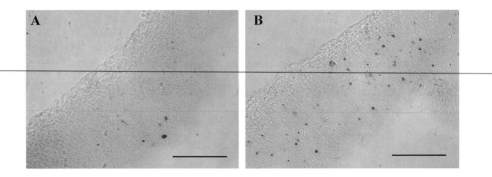

Fig. 1. TUNEL staining of fetal mouse brains (embryonic day 12.5) 24 hr after treatment with BrdU. **A**, control. **B**, BrdU. Sections including the neocortex were processed for this staining. Marked TUNEL-positive reactivity was observed in brains treated with BrdU (**B**). *Bars*, 100 μm

more directly. Consistent with the rat study, the present findings suggest that neurogenesis is toxicologically heterogeneous with respect to BrdU-induced toxicity during early development. Further, the present results indicate possible misestimation in previous stem cell studies using BrdU.

References

Kuwagata M, Muneoka KT, Ogawa T, Takigawa M, Nagao T (2004) Locomotor hyperactivity following prenatal exposure to 5-bromo-2'-deoxyuridine : Neurochemical and behavioral evidence of dopaminergic and serotonergic alterations Tox Lett 152:63-71

Miller MW, Nowakowski RS (1988) Use of bromodeoxyuridine- immunohistochemistry to examine the proliferation, migration and time of origin of cells in the central nervous system. Brain Res 457:44-52

Morris SM (1991) The genetic toxicology of 5-bromodeoxyuridine in mammalian cells. Mutat Res 258:161-188

Ogawa T, Nakamachi T, Ohtaki H, Hashimoto H, Shintani N, Baba A, Watanabe J, Kikuyama S, Shioda S (2005a) Monoaminergic neuronal development is not affected in PACAP-gene-deficient mice. Regul Pept 126:103-8

Ogawa T, Kuwagata M, Muneoka K, Shioda S (2005b) Neuropathological examination of fetal rat brain in the 5-bromo-2'-deoxyuridine-induced neurodevelopmental disorder model. Cong Anom 45:14-20

Soriano E, Del Rio JA (1991) Simultaneous immunocytochemical visualization of bromodeoxyuridine and neural tissue antigens. J Histochem Cytochem. 39: 255-263

Takahashi T, Goto T, Miyama S, Nowakowski RS, Caviness VS Jr (1999) Sequence of neuron origin and neocortical laminar fate: relation to cell cycle of origin in the developing murine cerebral wall. J Neurosci 19:10357-10371

Part III
Brain Functions in Health and Disease

Hypocretin-Deficient Narcolepsy as a Disease Model to Study the Hypothalamic Function in Health and Disease

Seiji Nishino

Center for Narcolepsy, Stanford University, 701B Welch Rd., Palo Alto, CA 94304, USA

Summary. Using forward (i.e. positional cloning in canine narcolepsy) and reverse (i.e. mouse gene knockout) genetics, genes involved in the pathogenesis of narcolepsy (hypocretin/orexin ligand and its receptor) in animals have been identified. Mutations in hypocretin related-genes are rare in humans, but hypocretin-ligand deficiency is found in many cases. This discovery is likely to lead to the development of new diagnostic tests and treatments. Hypocretins/orexins are novel hypothalamic neuropetides also involved in various hypothalamic functions such as energy homeostasis, autonomic and neuroendocrine functions. Hypocretin-deficient narcolepsy thus appears now to be a more complex condition than a simple sleep disorder, and it may serve a disease model to study the most fundamental hypothalamic function.

Key words. Hypocretin, orexin, narcolepsy, hypothalamus, energy homeostasis

1 Discovery of Narcolepsy Genes in Animals and Hypocretin/orexin Ligand Deficiency in Human Narcolepsy

Narcolepsy is a chronic disabling sleep disorder affecting 1:2000 individuals. Patients with narcolepsy suffer from excessive daytime sleepiness (EDS), cataplexy (sudden loss of muscle tone with emotional excitement), sleep paralysis and hypnagogic hallucination (Nishino and Mignot 1997). The patients could not maintain long bouts of wakefulness and sleep, and thus exhibit EDS during daytime and insomnia at night. Cataplexy, sleep paralysis and hypnagogic hallucination are often regarded as dissociated manifestations of rapid eye movement (REM) sleep phenomena due to the rapid transitions to REM Sleep during active wakefulness or at sleep onset (Nishino and Mignot 1997). Narcolepsy has been described in several other species including dogs, and recently in mice. Canine narcolepsy is naturally-occurring animal model, and both sporadic (in 17 breed) and familial forms (in Dobermans, Labrador and Dachshund) of narcolepsy exist. About 95% of human narcolepsy occurs sporadically, while 5% are familial cases (Nishino and Mignot 1997). Human narcolepsy is thought to be caused by interplay of genetic and environmental factors. It shows a tight association with human leukocyte antigen markers HLA-DR2 and HLA-DQB1*0602 (Nishino and Mignot 1997). Although a number of autoimmune diseases, such as multiple scoreless and insulin-dependent diabetes mellitus, is associated with certain HLA haplotypes, no direct evidence of the autoimmune involvement in the etiology of narcolepsy has been demonstrated. Narcolepsy was first described in the medical literature at the end of the 19th century. However, its pathophysiology was only elucidated at the end of the 20th century.

Using forward (i.e. positional cloning) and reverse genetics (i.e. gene targeting), two groups independently revealed the pathogenesis of narcolepsy in animals. Either the lack of the hypothalamic neuropeptide hypocretin/orexin ligand (mouse gene knockout) (Chemelli et al 1999) or mutations in one (hypocretin receptor 2 [*hcrtr 2*]) of the two hypocretin/orexin receptor genes (autosomal recessive canine narcolepsy) (Lin et al 1999) was observed to result in the narcolepsy phenotype. After extensive mutation screening, especially in familial and early-onset human narcolepsy, it has been demonstrated that mutations in the hypocretin-related gene is rare: only one early-onset (6 months' age) narcolepsy case has been described to date, caused by a single point mutation in the pre-prohypocretin gene (Peyron et al 2000). Despite the lack of genetic abnormalities in the hypocretin system, the large majority (85-90%) of patients with narcolepsy-cataplexy have low or undetectable hypocretin 1 ligand in their cerebrospinal fluid (CSF) (Nishino et al 2001b). This hypocretin deficiency is tightly associated with occurrence of cataplexy and HLA-DQ1*0602 positivity (Mignot et al 2002). Postmortem human studies, although few in the number of cases studied have confirmed hypocretin ligand deficiency (both hypocretin 1 and 2) in the narcoleptic brains (Peyron et al 2000; Thannickal et al 2000). Hypocretin deficiency was also observed in sporadic cases of canine narcolepsy (7 out of 7 studied, the result of 4 cases are reported in Ripley et al 2001), suggesting that the pathophysiogy of these animals mirror the most human cases. These discoveries, representing the most striking progress in narcolepsy research in the last 100 years, were only made within the last 4 years.

Low CSF hypocretin 1 levels are very specific to narcolepsy compared to other sleep and neurological disorders (Mignot et al 2002; Ripley et al 2001). The establishment of CSF hypocretin 1 measurement as a new diagnostic tool for human narcolepsy is therefore encouraging. Indeed, low

CSF hypocretin 1 levels (< 110 pg/ml) are being considered as supportive data for the diagnosis of narcolepsy-cataplexy in the 2nd revision of International Classification of Sleep Disorders (ICSD). Previously, no specific and sensitive diagnostic test for narcolepsy based on the pathophysiology of the disease was available, and the final diagnosis was often delayed for several years after the disease onset that typically occurs in adolescence. Many patients with narcolepsy and related EDS disorders are therefore likely to obtain immediate benefit from this new specific diagnostic test. Since the lack of hypocretin ligand is the major pathophysiology in most human subjects, replacements of hypocretin agonists may be a promising approach for treatment. In this respect, developments of small-molecular and centrally penetrable (i.e. non-peptide) hypocretin agonists are likely to be necessary. If hypocretin receptors are functioning even years after the disease onset, and hypocretin replacement therapy may be effective for most ligand-deficient narcolepsy. Cell transplantation, using such as the embryonic hypothalamus or neural stem cells, and gene therapy (prepro-hypocretin gene transfer using various vectors) might also be used to cure the disease in the future. The causes/mechanisms of the ligand deficiency in human narcolepsy remain unknown, but it is likely to be an acquired cell death of the hypocretin neurons (see Thannickal et al, 2000). This hypothesis is supported by two observations; (1) onset of the most sporadic cases of human narcolepsy (peripubety) is later than that of narcolepsy caused by genetic abbreviation of hypocretin ligands (prepuberty onsets of the single case of preprohypocretin gene mutation case, hcrtr 2-mutated narcoleptic dogs and preprohypocretin gene knockout mice) and (2) the postnatal abbreviation of hypocretin neurons induces narcolepsy phenotype more resemble (vs. hypocretin ligand knockout) to human narcolepsy phenotype (i.e. existence of obesity) (Hara et al 2001). If this is the case, the mechanisms of the hypocretin cell death should be determined to prevent and/or rescue the disease.

2 New Hypothalamic Neuropetides, Hypocretins/Orexins

The hypocretin/orexin system was recently discovered in 1998; only one year before the discovery of the narcolepsy genes. Two independent groups discovered the hypocretin/orexin system. The Scripps group (Stanford, Yale, Oslo Universities and Jackson Laboratory) discovered the hypocretins (1 and 2) by searching for mRNAs specifically expressed within the hypothalamus (i.e. subtractive polymerase chain reaction) (De Lecea et al 1998). The University Texas Southwestern Medical Center and Smith-

Kline Beechem group simultaneously discovered the same peptides, by searching the endogenous ligands for orphan G protein coupled receptors (Sakurai et al 1998). Orphan receptors are receptors whose sequence/structure is known, but their endogenous ligands are unknown. The Texas group named these new peptides orexins (A and B) after the Greek word for appetite since they found that central administration of orexins potently increased food intake in rats (Sakurai et al 1998). Hypocretin 1 and 2 are produced by cleavage of a single precursor peptide, preprohypocretin. Mammalian hypocretin 1 is 33 amino acids long with two intrachain disulfide bonds, whereas hypocretin 2 is a linear 28 amino acid peptide with C-terminal amidation. The amino acid sequences of both peptides are almost identical among examined mammalian species, and likely to be conserved among vertebrates. The combination of the discovery of the hypocretins with the identification of mutations in the hypocretin receptor as the cause of canine narcolepsy required the application of modern technologies in molecular biology and reverse pharmacology (i.e. isolation of physiological ligands for orphan receptors). These discoveries were indeed timely accelerating research into the hypocretin system.

3 Hypocretin/orexin System and Sleep Regulation

The discovery of narcolepsy genes in 1999 was also revolutionary for the basic sleep research, since typical approaches for basic sleep research were using classical anatomical, physiological and pharmacological techniques (McCarley 1995). The most current popular models for the regulation of sleep and wakefulness are constituted by monoaminergic, cholinergic and excitatory and inhibitory amino acid systems, although the direct involvement of other neuromodulators, such as adenosine and other humoral factors or autacoids (cytokines and prostaglandins) are suggested by some sleep researchers (see McCarley 1995). Using the canine model of narcolepsy, we have also intensively studied the roles of monoamines and acetylcholine in narcolepsy, partially due to the fact that all compounds currently used for the treatment of narcolepsy enhance monoaminergic tonus (dopamine and norepinephrine) and that monoamine and acetylcholine systems interact reciprocally for the control of REM sleep (Nishino and Mignot 1997; McCarley 1995).

Molecular pharmacological and genetic engineering approaches have also been recently used in the basic sleep research. However, most have also focused on theses known neurotransmitter systems implicated in sleep regulation. Therefore, the discovery of the narcolepsy genes was radical for the sleep community, although the involvement of hypocretin system for the control of vigilance had been speculated based on its anatomical projections: hypocretin neurons project to the most brain structures, such as the cortex, anterior and posterior hypothalamus, thalamus, and the brainstem monoaminergic and cholinergic nuclei, thought to be important for the sleep regulation (Peyron et al 1998). Many sleep researches subsequently focused on the hypocretin system, and some started to believe that hypocretins/orexins are key wake-promoting and REM sleep suppressing substances. Indeed, a series of studies have now proven that hypocretin system is the major excitatory neuromodulatory system that control activities of monoaminergic (dopamine, norepinephrine, serotonin and histamine) and cholinergic systems to control vigilance states (see Willie et al 2001). Thus it is likely that deficient in hypocretin neurotransmission induce the imbalance of these neurotransmitter systems. Indeed, dopamine and/or norepinephrine contents had been reported to be high in several brain structures in narcoleptic Dobermans and in human postmortem brains (see Nishino and Mignot 1997), possibly due to the compensatory mechanisms. These up-regulations alone are however, not enough to compensate sleep abnormalities in narcolepsy, and pharmacological treatments that enhance dopaminergic neurotransmission, such as amphetamine-like stimulant and modafinil (for EDS) and norepinephrine neurotransmission, such as noradrenaline uptake blockers (for cataplexy) are needed (Nishino and Mignot 1997). Histamine is another monoamine implicated in the control vigilance, and the histaminergic system is also likely to mediates the wake-promoting effect of hypocretin (Huang et al 2001). Interestingly, brain histamine contents both in hcrtr-2 mutated and ligand-deficient narcoleptic dogs are dramatically reduced (Nishino et al 2001a) and a preliminary result suggest the decreased histamine content in the CSF of human narcolepsy (Nishino et al 2002). Thus, compounds that enhance central histaminergic neurotransmission (such as histamine autoreceptor H3 antagonists), may be an interesting choice for the treatment of narcolepsy.

4 Hypothalamic Feeding Network System

The hypothalamus has long been implicated in the regulation of food intake, body mass, and energy balance (see, Schwartz et al 2000 and Willie et al 2001). The lateral hypothalamus is specifically responsible for the initiation of food intake, while the basomedial hypothalamic nuclei is believed to be associated with the satiation of food intake. Appetite-stimulating (orexigenic) neuropeptides, such as melanin concentrating hormone (MCH), galanin, and dynorphin have been reported to locate in the lateral hypothalamic area (Schwartz et al 2000). Leptin is another important molecule for the regulation of food intake, secreted by adipocytes in accordance with total body adipose mass and mediate the signal to the central and reduce food intake through the hypothalamic feeding net work system (Schwartz et al 2000). The arcuate nucleus in the hypothalamus is believed to be a major site of leptin-action for the suppression of food intake. Both orexigenic neuropeptide Y/agouti-related protein-coexpressing neurons and anorectic proopiomelanocortin/cocaine-amphetamine-regulated transcript-coexpressing neurons exist in the arcuate nucleus, and it is thought that leptin inhibits and excites these orexinergic and anorexinergic neurons, repetitively. Since hypocretin neurons locate exclusively in the lateral hypothalamic area and central administration of hypocretin 1 stimulates the food intake, many researchers started to study the role of hypocretin/orexin in feeding regulation (see Willie et al 2001).

Further anatomical and functional studies of hypocretin system for energy homeostasis has successfully delineated the roles of hypocretins in the hypothalamus feeding network (see Willie et al 2001). Hypocretin neurons densely project to the arcuate nucleus, and hypocretin inhibits anorexinergic and excites orexinergic arcuate neurons that are opposite to the effects found for leptin. Taken together with the fact that hypocretin antibody that neutralize endogenous hypocretin peptide reduces spontaneous food intake, hypocretin system likely to play physiological roles for the food intake as one of the important components of the hypothalamic feeding net work system (see, Willie et al 2001).

5 Narcolepsy as a Model for Studying the Physiological Roles of Hypocretin/Orexin System in Health and Disease

However, multiple peripheral and central systems are involved in the regulation of food intake, and that the potency of hypocretins to enhance food intake in animals was less striking than other well-studied hypothalamic peptides, such as neuropeptide Y. The functional importance of the hypocretin/orexin system in the feeding network system remains to be studied.

Interestingly, narcolepsy was reported to be associated with changes in energy homeostasis, even several decades ago. Narcolepsy patients are frequently (1) obese, (2) more often have insulin-resistant diabetes mellitus, (3) exhibit reduced food intake and (4) have lower blood pressure and temperature (see Nishino et al 2001b). Altered neuroendocrine functions in narcolepsy have also been reported in narcolepsy (Lammers et al 2005).

These findings however, have not received much attention, since they could be secondary to sleepiness or inactivity during daytime and to the altered sleep wake pattern. It is however, by our and others' data in humans (Nishino et al 2001b; Overeem et al 2002) and in the animal model (Hara et al 2001) that hypocretin deficiency is more directly involved in altered energy homeostasis in hypocretin deficient narcolepsy.

Two recent studies in animals further suggest that the hypocretin tonus may be involved in the behavioral and vigilance changes during long-term food restriction. Several authors previously reported that long-term food deprivation increase wakefulness and reduce REM sleep during the rest period (Dewasmes et al., 1989). We have recently observed that basal hypocretin tonus (measured by the mean day hypocretin levels by microdialysis experiments) increased with the long-term food deprivation, while the basic diurnal fluctuation pattern was preserved (Yoshida et al., unpublished observation). It is also demonstrated in mice that an increase in activity or wakefulness during resting period were typically observed in wild-type mice, but not observed in hypocretin-deficient narcoleptic mice (Beuckmann et al 2002). These results therefore suggest that hypocretin tonus may also involved in the adaptation during the long-term food restriction. It is also interesting that the neurotransmitter system, which regulates the hypocretins, is limited to feeding regulatory peptides of peripheral origin, such as ghrelin, leptin or changes in glucose utilization (Yamanaka et al 2002). Adaptation under restricted food is a very fundamental, and essential regulatory mechanism for survival. It is therefore likely that sleep and energy homeostasis under various conditions are coordinated by the hypocretin/orexin system. In this respect, narcolepsy is not only a chronic sleep disorder, it is an important model to study the fundamental hypothalamic function. Currently, it is not known how other feeding-related peptides are involved in vigilance control. Similarly, almost all models currently available for sleep regulation do not incorporate the influence of food intake or food availability on vigilance control. A series of these discoveries also deserve further study on the fundamental links between sleep regulation and energy homeostasis and/or feeding.

The hypocretin system may also be involved in rewarding/reinforcement as well as stimulant abuse. We are especially interested in the results of the hypocretin system in amphetemine abuse. The abuse of amphetamine-like psychostimulants is a serious and growing problem in many countries.

Amphetamine-like psychostimulants have been prescribed for the treatment of narcolepsy for over 60 years. Considering the prevalence of narcolepsy (1/2000) and the fact that about 95% of narcoleptic patients are currently treated with pharmacological compounds, the total cumulative number of narcoleptic patients receiving psychostimulants is large. Nevertheless, psychostimulant abuse in narcoleptic subjects is extremely rare. It is reported that hypocretin-1 administration in rats induces hyperlocomotion, stereotypy possibly by exciting the midbrain dopamine neurons

(Nakamura et al 2000), the brain structure important for rewarding and stimulant abuse. We therefore hypothesized that hypocertin deficient-narcolepsy may be resistant to psychostimulant abuse.

To experimentally determine whether hypocretin/orexin-deficient narcolepsy is resistant to amphetamine abuse, we compared amphetamine-induced locomotor sensitization between preproporexin knockout and their littermate wild type mice, and between orexin/ataxin-3 transgenic narcoleptic and their littermate wild type mice. The locomotor sensitization experiments are typically used to evaluate the susceptibility to stimulant abuses. We found that both preproporexin knockout and orexin/ataxin-3 transgenic narcoleptic mice are less responsive in developing amphetamine-induced locomotor sensitization. Further investigation of amphetamine reward responses (behavioral sensitization, conditioned place preference test) in hypocretin-deficient mice is thus warranted.

6 Conclusion

Although the major pathphysiology of human narcolepsy (i.e., hypocretin/orexin ligand deficiency) is revealed, narcolepsy research is not the end and is rather the beginning of a new era. New treatments based on new findings, such as hypocretin replacements/gene therapy, needs to be developed. The mechanisms of the hypocretin deficiency should be determined to prevent and/or rescue the disease.

The hypocretin/orexin system is involved in various fundamental hypothalamice functions, such as vigilance control, energy homeostasis, neuroendcrine, and autonomic nerve functions. A series of experiments also pointed out that sleep and energy homeostasis under various conditions are coordinated by the hypocretin/orexin system. Altered energy homeostasis and neuroendocrine functions are also reported in human narcolepsy.

It thus appears that Narcolepsy is not only a chronic sleep disorder, but it is also an important model for studying the fundamental hypothalamic functions in health and disease.

References

Alam MN, Gong H, Alam T, Jaganath R, McGinty D, Szymusiak R (2002): Sleep-waking discharge patterns of neurons recorded in the rat perifornical lateral hypothalamic area. J Physiol 538:619-31

Aston-Jones G, Chen S, Zhu Y, Oshinsky ML (2001): A neural circuit for circadian regulation of arousal. *Nat Neurosci* 4:732-8

Beuckmann CT, Willie JT, Hara J, Yamanaka K, Sakurai T, Yanagisawa M (2002): Orexin neuron-ablated mice fail to increase vigilance and locomotor activity in response to fasting. *Sleep* 25, Abstract suppliment:A353-4

Bourgin P, Huitron-Resendiz S, Spier AD, et al (2000): Hypocretin-1 modulates rapid eye movement sleep through activation of locus coeruleus neurons. *J Neurosci* 20:7760-5

Broughton R, Dunham W, Lutley K, Newman J (1987): Ambulatory 24 hour sleep-wake monitoring in narcolepsy-cataplexy compared to matched control. *Electroenceph Clin Neurophysiol* 70:473-481

Chemelli RM, Willie JT, Sinton CM, et al (1999): Narcolepsy in orexin knockout mice: molecular genetics of sleep regulation. *Cell* 98:437-451

Chou TC, Lee CE, Lu J, et al (2001): Orexin (hypocretin) neurons contain dynorphin. *J Neurosci* 21:RC168

De Lecea L, Kilduff TS, Peyron C, et al (1998): The hypocretins: Hypothalamus-specific peptides with neuroexcitatory activity. *Proc Natl Acad Sci USA* 95:322-327

Dewasmes G, Duchamp C, Minaire Y(1989). Sleep changes in fasting rats.Physiol Behav 46:179-84

Estabrooke IV, McCarthy MT, Ko E, et al (2001): Fos expression in orexin neurons varies with behavioral state. *J Neurosci* 21:1656-62

Franken P, Tobler, I. Borbély A (1991): Sleep homeostasis in the rat: simulation of the time course of EEG slow wave activity. Neuroscience Letters 130:141-144

Hara J, Beuckmann CT, Nambu T, et al (2001): Genetic ablation of orexin neurons in mice results in narcolepsy, hypophagia, and obesity. *Neuron* 30:345-54

Huang ZL, Qu WM, Li WD, et al (2001): Arousal effect of orexin A depends on activation of the histaminergic system. *Proc Natl Acad Sci U S A* 98:9965-70

Kiyashchenko LI, Mileykovskiy BY, Maidment N, et al (2002): Release of hypocretin (orexin) during waking and sleep states. *Journal of Neuroscience* 22:5282-6

Lammers GL, Overeem S, Pijil H (2005): Neuroendocrinology of human narcolepsy. In Nishino S, Sakurai T (eds), *The Orexin/Hypocretin System: Its Physiology And Pathophysiology*. Totowa, New Jersey: Humana

Lin L, Faraco J, Li R, et al (1999): The sleep disorder canine narcolepsy is caused by a mutation in the hypocretin (orexin) receptor 2 gene. *Cell* 98:365-76

McCarley R (1995): Sleep, dreams, and states of consciousness. In Conn PM (ed), *Neuroscience in Medicine*. Philadelphia: J.B. Lippincott Company, pp 537-553

Mignot E, Lammers GJ, Ripley B, et al (2002): The role of cerebrospinal fluid hypocretin measurement in the diagnosis of narcolepsy and other hypersomnias. *Arch Neurol* 59:1553-62

Mitler MM, Dement WC (1977): Sleep studies on canine narcolepsy: pattern and cycle comparisons between affected and normal dogs. *Electroencephalogr Clin Neurophysiol* 43:691-9

Nakamura T, Uramura K, Nambu T, et al (2000): Orexin-induced hyperlocomotion and stereotypy are mediated by the dopaminergic system. *Brain Res* 873:181-7

Nishino S, Fujiki N, Ripley B, et al (2001a): Decreased brain histamine contents in hypocretin/orexin receptor-2 mutated narcoleptic dogs. *Neurosci Lett*

313:125-8

Nishino S, Mignot E (1997): Pharmacological aspects of human and canine narcolepsy. *Prog Neurobiol* 52:27-78

Nishino S, Riehl J, Hong J, Kwan M, Reid M, Mignot E (2000): Is narcolepsy REM sleep disorder? Analysis of sleep abnormalities in narcoleptic Dobermans. *Neuroscience Research* 38:437-446

Nishino S, Ripley B, Mignot E, Benson KL, Zarcone VP (2002): CSF hypocretin-1 levels in schizophrenics and controls: relationship to sleep architecture. *Psychiatry Res* 110:1-7

Nishino S, Ripley B, Overeem S, et al (2001b): Low CSF hypocretin (orexin) and altered energy homeostasis in human narcolepsy. *Ann Neurol* 50:381-388

Overeem S, Dalmau J, Bataller L, et al (2001): Secondary narcolepsy in patients with praneoplastic anti-Ma2 antibodies is associated with hypocretin deficiency. *J Sleep Res* 11 (suppl. 1):166-167

Overeem S, Koh SW, Pijl H, Lammers GJ, Meinders AE (2002): Body weight and -composition in patients with narcolepsy versus idiopathic hypersomnia. *Sleep* 25 Abstract Suppliment:A79

Pedrazzoli M, Ling L, D'Almeida V, et al (2002): Hypocretin levels in rat CSF after paradoxical (REM) sleep deprivation. *J Sleep Res* (suppl.) 1:171

Peyron C, Faraco J, Rogers W, et al (2000): A mutation in a case of early onset narcolepsy and a generalized absence of hypocretin peptides in human narcoleptic brains. *Nat Med* 6:991-7

Peyron C, Tighe DK, van den Pol AN, et al (1998): Neurons containing hypocretin (orexin) project to multiple neuronal systems. *J Neurosci* 18:9996-10015

Ripley B, Overeem S, Fujiki N, et al (2001): CSF hypocretin/orexin levels in narcolepsy and other neurological conditions. *Neurology* 57:2253-8

Sakurai T, Amemiya A, Ishil M, et al (1998): Orexins and orexin receptors: a family of hypothalamic neuropeptides and G protein-coupled receptors that regulate feeding behavior. *Cell* 92:573-585

Salomon RM, Ripley B, Kennedy J, et al (2002): Diurnal Variation of CSF hypocretin-1 (Orexin A) levels in control and depressed subjects. *Sleep* 25 Abstract Suppliment:A12

Schwartz MW, Woods SC, Porte D, Jr., Seeley RJ, Baskin DG (2000): Central nervous system control of food intake. *Nature* 404:661-71

Tafti M, Rondouin G, Besset A, Billiard M (1992) Sleep deprivation in narcoleptic subjects: effect on sleep stages and EEG power density. *Electroencephalogr Clin Neurophysiol* 83:339-49

Thannickal TC, Moore RY, Nienhuis R, et al (2000): Reduced number of hypocretin neurons in human narcolepsy. *Neuron* 27:469-74

Yamanaka A, Hara J, Tsujino N, Beuckmann C, Yanagisawa M, Sakurai T (2002): Regulation of orexin neurons by peripheral nutritional signals: roles of leptin, ghrelin and glucose. *Sleep* 25, Abstract Suppliment:A356-7

Yoshida Y, Fujiki N, Nakajima T, et al (2001): Fluctuation of extracellular hypocretin-1 (orexin A) levels in the rat in relation to the light-dark cycle and sleep-wake activities. *Eur J Neurosci* 14:1075-81

Willie JT, Chemelli RM, Sinton CM, Yanagisawa M (2001): To eat or to sleep?

Orexin in the regulation of feeding and wakefulness. *Annu Rev Neurosci* 24:429-58

The Expression of Synaptic Vesicle Proteins after Chronic Antidepressant Treatment in Rat Brain

Misa Yamada[1,2], Kou Takahashi[1,2], Chika, Kurahashi[2], Mitsuhiko Yamada[1], and Kazuo Honda[2]

[1]Department of Psychogeriatrics, National Institute of Mental Health, National Center of Neurology and Psychiatry, 4-1-1 Ogawahigashimachi, Kodaira, Tokyo 187-8502, Japan
[2]Departments of Pharmacology, School of Pharmaceutical Sciences, Showa University, 1-5-8 Hatanodai, Shinagawa, Tokyo 142-8555, Japan

Summary. The biological basis for the therapeutic mechanisms of depression are still unknown. We have previously performed EST analysis and identified some common biological changes induced after chronic antidepressant treatment as antidepressant related genes/ESTs : ADRG#1-707. Then, we developed our original cDNA microarray on which ADRG#1-707 were spotted, for rapid secondary screening of candidate genes as the novel therapeutic targets. With this microarray, we found that the expression of some of the ADRGs were related to neurotransmiter release and located on synaptic vesicle. Indeed, VAMP2/synaptobrevin, cysteine string protein, synapsin I, Rab-1A and Rab-3B were induced after chronic sertraline treatment in rat frontal cortex. Western blot analysis also demonstrated the induction of these ADRGs at protein levels after chronic treatment with imipramine and sertraline. In addition, synaptophysin and secretogranin II, often used as a marker protein for small synaptic vesicle or large dense core granule were significantly increased after chronically treatment with antidepressants. On the other hand, the expression of SNAP-25 and syntaxin-1, which are used as markers for synapse and make a SNARE-complex with VAMP2, were not affected by these treatments. These results suggested that the number of synaptic vesicles, but not the number of synapses, was increased after chronic antidepressant treatment. The synaptic vesicles and proteins may be a new target molecular system for antidepressant.

Key words. Depression, antidepressant, microarray, synaptic vesicle, SNARE complex

1 Introduction

It has been demonstrated that typical antidepressants acutely inhibit the monoamine reuptake in nerve terminals resulting in significant increase in synaptic concentrations of monoamines, noradrenaline or serotonin. However, there is a latency period of several weeks before the onset of clinical effect of antidepressants. There are several preclinical investigation shown the delayed action of antidepressants on mood, motivation and cognition is not linked to their primary mechanism of action but rather to the development of various modifications (Duman and Vaidya 1998). Hyman and Nestler proposed an "initiation and adaptation" model to describe the drug-induced neural plasticity associated with the long-term actions of antidepressants in the brain (Hyman and Nestler 1996). However, the detailed mechanisms underlying such drug-induced adaptive neuronal changes are as of yet unknown. The delay of clinical effect from antidepressants could be the result of indirect regulation of neural signal transduction systems or changes at the molecular level by an action on gene transcription following chronic treatment. Indeed, there are selective effects of antidepressants on specific immediate early genes and transcription factors. These molecules activate or repress genes encoding specific proteins by binding to a regulating element of DNA. These functional proteins may be involved in critical steps in mediating treatment-induced neural plasticity. Therefore, we demonstrated that certain novel candidate genes and molecular systems may underlie the mechanism of action of antidepressants.

2 EST Analysis and Fabrication of the Original cDNA Microarray for Antidepressant Research

We have performed expressed-sequence tag (EST) analysis to identify some common biological changes induced after chronic treatment of two different classes of antidepressants, imipramine (a tricyclic antidepressant) and sertraline (a serotonin selective reuptake inhibitor, SSRI). Identification of quantitative changes in gene expression that occur in the brain after chronic antidepressant treatment can yield novel molecular machinery responsible for therapeutic effect of antidepressant. Until now, we have molecularly cloned 707 cDNA fragments which were named them antidepressant related genes, ADRG from rat frontal cortex, hippocampus and hipothalamus (Yamada et al. 1999). More recently, for high throughput secondary screening of candidate genes, each of the ADRGs were spotted in duplicate onto glass slides to develop our original microarray, ADRG microarray. After hybridization with samples obtained from sertraline treated rat frontal cortex and normalization of the signals for both negative and positive controls, we have identified several interesting candidate genes and ESTs on the ADRG microarray that showing increased or decreased expression compared from control.

3 New Candidate Molecular Systems in Depression Research

We found some of the candidate molecules and molecular systems with this ADRG microarray. Interestingly, the expression of some of the ADRGs related to neurotransmiter release and located on synaptic vesicles were induced after chronic treatment with sertraline in rat frontal cortex (Fig. 1). We previously reported that the expression of ADRG55, identified as cysteine string protein (CSP), was induced after chronic antidepressant treatment (Yamada et al. 2001). CSP is localized to synaptic vesicle membranes and modulates the activity of presynaptic calcium channels, resulting in neurotransmitter release at the nerve terminal in the central nervous system (Gundersen et al. 1995). In addition, we have also demonstrated that the expression of ADRG14, identified as vesicle associated membrane protein VAMP2/ synaptobrevin, was induced after chronic antidepressant treatment (Yamada et al. 2002). VAMP2/ synaptobrevin is a key component of the synaptic vesicle docking/fusion machinery that forms the SNARE (soluble N-ethylmaleimide-sensitive fusion protein attachment protein receptor) complex (Weis and Scheller 1998). More recently, we identified some more ADRGs related to neurotransmitter release and located on synaptic vesicles, including synapsin I, Rab-1A and Rab-3B. Synapsin I is an actin-binding protein that localized on the cytoplasmic face of small synaptic vesicles and inhibits neurotransmitter release, an effect that is abolished upon its phosphorylation by Ca^{2+}/calmodulin-dependent protein kinase II. Rab protein is low molecular weight GTP-binding protein of the Ras superfamily of GTPases. Rab protein is involved in intracellular membrane fusion reactions located in cytoplasmic face of organelles and vesicles. The synapsin I, Rab-1A and Rab-3B were induced after chronic antidepressant treatment in rat frontal cortex, when determined by ADRG microarray.

Western blot analysis also demonstrated the induction of these synaptic vesicle proteins after chronic treatment with imipramine and sertraline in rat frontal cortex. These results indicate that two possibilities i) the number of synaptic vesicles is increased, ii) the number of synapses is increased after chronic antidepressant treatment. To investigate the first possibility, the expression of synaptophysin and secretogranin II were determined by Western blot analysis. Then, to investigate the second possibility, the expression of SNAP-25 and syntaxin-1, which are used as markers for synapse and make a SNARE-complex with VAMP2, were determined. Interestingly, the expression of both synaptophysin (a marker protein for small synaptic vesicles) and secretogranin II (a marker protein for large dense core granules) were significantly increased after chronic treatment with antidepressants. On the other hand, the expression of SNAP-25 and syntaxin-1 were unaffected by these treatments. These results strongly suggested that the number of synaptic vesicles, but not the number of synapses, was increased after chronic antidepressant treatment.

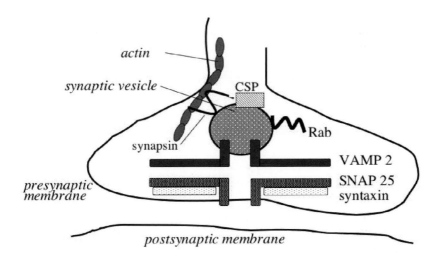

Fig. 1. Synaptic vesicle proteins identified as ADRG and related proteins. The expression of some of the ADRGs related to neurotransmiter release and located on synaptic vesicle including VAMP2/synaptobrevin, cysteine string protein (*CSP*), synapsin I, Rab-1A and Rab-3B were induced after chronic sertraline treatment in rat frontal cortex determined by ADRG microarray

There are several articles that reporting the change of synaptic protein expression, activity or phosphorylation affected by antidepressant treatments (Popoli et al. 1995). Further, long-term treatment of hippocampal slice cultures with brain-derived neurotrophic factor (BDNF) that is one of the key molecule induced by antidepressant treatment (D'Sa and Duman 2002) increased the number of docked vesicles, but not that of reserve pool vesicles at CA1 excitatory synapses. BDNF also increased the levels of vesicle proteins synaptophysin, synaptobrevin, and synaptotagmin, without affecting the presynaptic membrane proteins syntaxin and SNAP-25, or the vesicle-binding protein synapsin-I (Tartaglia et al. 2001). Taken together, these finding may suggest a link between the modulation of synaptic vesicle proteins and the therapeutic mechanisms of antidepressants.

4 Conclusion

In the present study, we have demonstrated that the synaptic vesicles and proteins may play a role in the therapeautic molecular systems of antidepressant treatment. These alterations of the expression pattern of synaptic vesicle proteins may also be associated with neural plasticity including modifications in neural connectivity, and modulation of synaptic vesicle density that occur during antidepressant treatment. Here, we propose that the changes in neural plasticity are implicated in the adaptive mechanisms underlie the delayed onset of therapeutic action of antidepressants. Our results may contribute to a novel model for the therapeutic mechanism of depression and new molecular targets for the development of therapeutic agents.

References

D'Sa C, Duman RS (2002) Antidepressants and neuroplasticity. Bipolar Disord 4: 183-194

Duman RS, Vaidya VA (1998) Molecular and cellular actions of chronic electro-convulsive seizures. J Ect 14: 181-193

Gundersen CB, Mastrogiacomo A, Umbach JA (1995) Cysteine-string proteins as templates for membrane fusion: models of synaptic vesicle exocytosis. J Theor Biol 172: 269-277

Hyman SE, Nestler EJ (1996) Initiation and adaptation: a paradigm for understanding psychotropic drug action. Am J Psychiatry 153: 151-162

Popoli M, Vocaturo C, Perez J, Smeraldi E, Racagni G (1995) Presynaptic Ca2+/calmodulin-dependent protein kinase II: autophosphorylation and activity increase in the hippocampus after long-term blockade of serotonin reuptake. Mol Pharmacol 48: 623-629

Tartaglia N, Du J,Tyler WJ, Neale E, Pozzo-Miller L, Lu B (2001) Protein synthesis-dependent and -independent regulation of hippocampal synapses by brain-derived neurotrophic factor. J Biol Chem 276: 37585-37593

Weis WI, Scheller RH (1998) Membrane fusion. SNARE the rod, coil the complex. Nature 395: 328-329

Yamada M, Yamada M, Kiuchi Y, Nara K, Kanda Y, Morinobu S, Momose K, Oguchi K, Kamijima K, Higuchi T (1999) Identification of a novel splice variant of heat shock cognate protein 70 after chronic antidepressant treatment in rat frontal cortex. Biochem Biophys Res Commun 261: 541-545

Yamada M, Takahashi K, Tsunoda M, Nishioka G, Kudo K, Ohata H, Kamijima K, Higuchi T, Momose K, Yamada M (2002) Differential expression of VAMP2/synaptobrevin-2 after antidepressant and electroconvulsive treatment in rat frontal cortex. Pharmacogenomics J 2: 377-382

Yamada M, Yamada M, Yamazaki S, Nara K, Kiuchi Y, Ozawa H, Yamada S, Oguchi K, Kamijima K, Higuchi T, Momose K (2001) Induction of cysteine string protein after chronic antidepressant treatment revealed by ADRG microarray. Neurosci Lett 301: 183-186

Identification of Molecular Systems Responsible for the Therapeutic Effect of Antidepressant

Mitsuhiko Yamada[1, 2] and Misa Yamada[1, 3]

[1]Department of Psychogeriatrics, National Institute of Mental Health, National Center of Neurology and Psychiatry, 4-1-1 Ogawahigashimachi, Kodaira, Tokyo 187-8502, Japan
[2]Departments of Psychiatry, School of Medicine, Showa University, 1-5-8 Hatanodai, Shinagawa, Tokyo 142-8666, Japan
[3]Departments of Pharmacology, School of Pharmaceutical Sciences, Showa University, 1-5-8 Hatanodai, Shinagawa, Tokyo 142-8555, Japan

Summary. Although blockade by antidepressants of monoamine uptake into nerve endings is one of the cornerstones of the monoamine hypothesis of depression, there is a clear discrepancy between the rapid effects of antidepressants in increasing synaptic concentrations of monoamine and the lack of immediate clinical efficiency of antidepressant treatment. Pharmacogenomics, functional genomics and proteomics are powerful tools that can be used to identify genes/ESTs or molecular systems affected by antidepressants. Using a differential cloning strategy, we and other groups have isolated genes that are differentially expressed in the brain after chronic antidepressant treatment. Some of these candidate genes may encode functional molecular systems or pathways induced by chronic antidepressant treatment. Defining the roles of these molecular systems in drug-induced neural plasticity is likely to transform the course of research on the biological basis of depression. Such detailed knowledge will have profound effects on the diagnosis, prevention, and treatment of depression.

Key words. Depression, antidepressant, differential cloning, system biology

1 Introduction

Depression is one of the major psychiatric diseases; represent abnormality of emotional, cognitive, autonomic and endocrine functions. Anti- depressants are very effective agents for the prevention and treatment of depression, and have been used clinically for more than 50 years. Although the therapeutic action of these antidepressants most likely involves the regulation of serotonergic and noradrenergic signal transduction pathways, to date, no consensus has been reached concerning the precise molecular and cellular mechanism of action of these drugs. Many antidepressants acutely regulate monoaminergic signal transduction within a few hours of initial treatment. However, at the same time, the onset of the clinical effect of these drugs lags by several weeks. A satisfying explanation for the discrepancy in the acute increase of synaptic monoamines and delayed clinical effect remains elusive. Consequently, the monoamine hypothesis does not fully explain this clear discrepancy. Novel biological approaches beyond the "monoamine hypothesis" are definitely expected to cause paradigm shifts in the future of depression research. In this article, we demonstrated that certain novel candidate molecular systems might underlie the mechanism of action of antidepressants.

2 The Delayed Clinical Effects and Changes in Gene Expression Elicited by Antidepressant in the Brain

To advance our understanding of the therapeutic actions of anti- depressants, we must now extend our efforts beyond theories based on the simple pharmacology of the synapse. This new effort must seek a deeper understanding of cellular and molecular neurobiology as well as examine the architecture and function of relevant neural systems. Many now believe that changes in brain gene expression, which are elicited after chronic antidepressant treatment, might underlie the drug-induced neural plasticity associated with the long-term actions of antidepressants in the brain and their clinical effects.

On the other hand, there are several preclinical investigations shown that the delay of clinical effect from antidepressants could be the result of indirect regulation of neural signal transduction systems or changes at the molecular level by an action on gene transcription following chronic treatment. Indeed, there are selective effects of antidepressants on specific immediate early genes and transcription factors including c-fos, zif268, NGFI-A and the phosphorylation of CRE binding protein. These molecules activate or repress genes encoding specific proteins by binding to a regulating element of DNA. These functional proteins may be involved in critical steps in mediating treatment-induced neural plasticity (see review by Yamada et al., 2002).

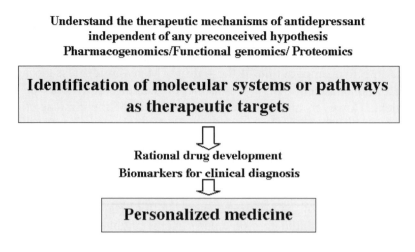

Fig. 1. Pharmacogenomics, Functional genomics and Proteomics are powerful tools that can be used to identify neuronal systems or pathways affected by anti-depressants

Recent developments in molecular neurobiology provide new conceptual and experimental tools to investigate, and facilitate understanding of the mechanisms by which antidepressants produce long-lasting alterations in brain function. The emerging techniques and powerful tools derived from the relatively new subfields of genomics and proteomics hold great promise for the identification of genes and gene products that are altered by chronic antidepressant treatment or other effective therapeutic manipulations, such as electroconvulsive treatment (ECT). Using a differential cloning strategy, we and other groups have isolated genes that are differentially expressed in the brain after chronic antidepressant treatment. Independent of any preconceived hypothesis, these genes and proteins have been implicated in a physiological or pathophysiological process. Defining the roles of the candidate systems in antidepressant-induced neural plasticity is likely to transform the course of research on the biological basis of mood disorders, leading to develop a personalized medicine (Fig.1). Such detailed knowledge will have profound effects on the diagnosis, prevention, and treatment of depression.

3 New Candidate Molecular Systems Responsible for the Therapeutic Effect of Antidepressant

Many now believe that changes in brain gene expression, which are elicited after chronic antidepressant treatment, might underlie the drug-induced neural plasticity associated with the long-term actions of antidepressants in the brain and their clinical effects. Here, we introduce three of the new candidate molecular systems in antidepressant research.

3.1 Adult Neurogenesis in the Hippocampus

Although depression involves many psychological and social factors, it also represents a biological process: the effects of repeated exposure to stress on a vulnerable brain. Preclinical and clinical research has focused on the interactions between stress and depression and their effects on the hippocampus (Duman et al., 1999). The hippocampus is one of several brain regions that, when exposed to stressful stimuli, can contribute to the emotional, cognitive, and vegetative abnormalities found in depressed patients. This region of the brain is also involved in the feedback regulation of the hypothalamus pituitary adrenal axis, the dysfunction of which is associated with depression. Recent studies suggest that stress-induced atrophy and loss of hippocampal neurons may contribute to the pathophysiology of depression. Interestingly, hippocampal volume is decreased in patients with stress-related psychiatric illnesses, including depression and post-traumatic stress disorder (Sapolsky and Duman, 2000; Sheline et al., 1996).

In vitro and in vivo data provide direct evidence that brain-derived neurotrophic factor (BDNF) is one of the key mediators of the therapeutic response to antidepressants (D'Sa and Duman, 2002). BDNF promotes the differentiation and survival of neurons during development and in the adult brain, as well as in cultured cells. Stress decreases the expression of BDNF, and reduced levels could contribute to the atrophy and compromised function of stress-vulnerable hippocampal neurons. In contrast, antidepressant treatment increases the expression of BDNF in the hippocampus, and could thereby reverse the stress-induced atrophy of neurons or protect these neurons from further damage (Duman, 1998; Duman et al., 1997). These findings have resulted in the development of a novel model of the mechanism of antidepressant action and have suggested new targets for the development of therapeutic agents.

While hippocampal volume can decrease in disease, the hippocampus is also one of only a few brain regions where the production of neurons normally occurs throughout the lifetime of several species of animals, including humans (Eriksson et al., 1998). Hippocampal neurogenesis is influenced by several environmental factors and stimuli (Gould and Tanapat, 1999; Nilsson et al., 1999; van Praag et al., 1999). For example, both acute and chronic stress cause decreases in cell proliferation. On the other hand, administration of several different classes of antidepressant, as opposed to non-antidepressant, agents increases the number of BrdU-labeled cells, indicating that this is a common and selective action of antidepressants (Malberg et al., 2000). In addition, recent evidence indicates that electroconvulsive seizures (an animal model of ECT in humans) can also enhance neurogenesis in rat hippocampus (Hellsten et al., 2002). These findings raise the possibility that increased cell proliferation and increased neuronal number may be a mechanism by which antidepressant treatment mitigates stress-induced atrophy and loss of hippocampal neurons, and thus may contribute to the therapeutic actions of antidepressant treatment. Furthermore, increased formation of new neurons in the hippocampus related to antidepressant treatment may lead to altered expression of genes specifically expressed in immature neurons. Therefore, observed changes in gene expression may reflect alterations in cell composition of the tissue rather than changes in individual neurons.

3.2 Vesicular Transport/Exocytotic Machinery

In our laboratory, we employed the RNA fingerprinting technique, a modified differential display PCR, to identify biochemical changes induced by chronic antidepressant treatments. To date, we have cloned several cDNA candidates as ESTs from rat frontal cortex and hippocampus. Some of these candidate cDNAs should be affected by antidepressants and are thus named antidepressant related genes (ADRGs). Among these ADRGs, we previously demonstrated that a unique cysteine-rich protein, called cysteine string protein (CSP), is clearly elevated in rat brain after chronic antidepressant treatment (Yamada et al., 2001). In rat brain, CSP interacts with VAMP2 in synaptic vesicle membranes and modulating the activity of presynaptic calcium channels, resulting in neurotransmitter release at the nerve terminal. Considerable evidence indicates that VAMP-2 is a key component of the synaptic vesicle transport/docking/fusion machinery that forms the SNARE (soluble N-ethylmaleimide-sensitive fusion protein attachment protein receptor) complex. Fusion of vesicles with the plasma membrane leads to exocytosis, which mediates the release of neurotransmitter into the synapse. Recently, we demonstrated a significant increase of both VAMP2 mRNA and protein levels in rat frontal cortex after chronic treatment with antidepressant and repeated ECT (Yamada et al., 2002). In this context, pharmacological modulations of CSP and VAMP2 expressions would also be predicted to alter neurotransmitter release. Interestingly, the work of others shows that acute and chronic administration of antidepressants diminishes the release of glutamate and aspartate, and inhibits veratridine-evoked 5-HT release (Golembiowska and Dziubina, 2000).

On the other hand, post-mortem depressive suicide brain samples were investigated to test the hypothesis that the regulation of SNARE proteins could be abnormal in depression (Honer et al., 2002). Interestingly, the immunoreactivity of VAMP2 was increased in depressive group. Further, the correlation between VAMP2 and other SNARE protein or synaptophysin were remarkably weak, and in some cases clearly non-significant. Of course, there were limitation of the availability of tissue for investigation and drug treatment history; the authors concluded that the abnormalities of SNARE complex could represent a molecular substrate for abnormalities of neural connectivity in depression.

Popoli and his fellows have demonstrated that the long-term treatment with antidepressants induced presynaptic CaM Kinase II activity, one of the kinases present involved in the modulation of transmitter release. Further, phosphorylation of synapsin I and synaptotagmin, the presynaptic substrates of CaM Kinase II were also increased after these treatments (Celano et al., 2003). In addition, in the amygdala of rats that received daily treatment with the TCA imipramine for 3 weeks, the gene encoding a mutation suppressor for the Sec4-8 yeast (Mss4) transcript was over-expressed (Andriamampandry et al., 2002). Mss4 protein has the properties of a guanine nucleotide exchange factor, and interacts with several members of the Rab family implicated in Ca^{2+}-dependent exocytosis of neurotransmitters. Interestingly, Mss4 transcripts were specifically down-regulated in the hippocampus and amygdala of rats after exposure to chronic, mild stress. These findings suggest that gene expression- dependent alterations of neuronal transmitter release may be an important component of the pharmacological action of antidepressants.

3.3 Axonal/Dendritic Outgrowth and Sprouting

Interestingly, vesicular docking/fusion at the plasma membrane is responsible not only for the release of neurotransmitters, but also for surface expression of plasma membrane proteins and lipids. Therefore, exocytosis plays a fundamental role in axonal/dendritic outgrowth and sprouting because both processes involve major increases in the surface area of the plasma membrane. In addition, treatment with chronic antidepressant increases the expression of GAP-43 in the rat dentate gyrus (Chen et al., 2003). Because GAP-43 regulates growth of axons and modulates the formation of new connections, these findings suggest that chronic antidepressant treatment may have an effect on structural neuronal plasticity in the central nervous system. As mentioned above, ECT is a safe and the most effective treatment for severely depressed patients who are resistant to antidepressant medications. Interestingly, the common effects of antidepressants and ECT on connectivity and synaptic plasticity in the dentate gyrus are likely to relate to affective functions of depression (Stewart and Reid, 2000). Consistent with these findings are data demonstrating that chronic electroconvulsive seizure administration in animals induces sprouting of the granule cell mossy fiber pathway in the hippocampus (Vaidya et al., 1999).

4 Conclusion

In this article, we demonstrated that certain novel candidate molecular systems or pathways might underlie the mechanism of action of antidepressants. Defining the roles of these molecular systems in drug-induced neural plasticity is likely to transform the course of research on the biological basis of depression. Identification of such targets will advance future efforts in the quest to develop effective therapeutics that have a new mode of action in the brain. Such detailed knowledge will have profound effects on the diagnosis, prevention, and treatment of depression. In conclusion, in the era of functional genomics, novel biological approaches beyond the "monoamine hypothesis" are expected to evoke paradigm shifts in the future of depression research. Additional work will be necessary to test this hypothesis.

References

Andriamampandry C, Muller C, Schmidt-Mutter C, Gobaille S, Spedding M, Aunis D, Maitre M (2002) Mss4 gene is up-regulated in rat brain after chronic treatment with antidepressant and down-regulated when rats are anhedonic. Mol Pharmacol 62: 1332-1338

Celano E, Tiraboschi E, Consogno E, D'Urso G, Mbakop MP, Gennarelli M, de Bartolomeis A, Racagni G, Popoli M. (2003) Selective regulation of presynaptic calcium/calmodulin-dependent kinase II by psychotropic drugs. Biol Psychiatry 53: 442-449

Chen B, Wang JF, Sun X, Young LT (2003) Regulation of GAP-43 expression by chronic desipramine treatment in rat cultured hippocampal cells, Biol Psychiatry 53: 530-537

D'Sa C, Duman RS (2002) Antidepressants and neuroplasticity. Bipolar Disord 4: 183-194

Duman RS (1998) Novel therapeutic approaches beyond the serotonin receptor. Biol Psychiatry 44: 324-335

Duman RS, Heninger GR, Nestler EJ (1997) A molecular and cellular theory of depression. Arch Gen Psychiatry 54: 597-606

Duman RS, Malberg J, Thome J (1999) Neural plasticity to stress and antidepressant treatment. Biol Psychiatry 46: 1181-1191

Eriksson PS, Perfilieva E, Bjork-Eriksson T, Alborn AM, Nordborg C, Peterson DA, Gage FH, Duman RS, Heninger GR, Nestler EJ (1998) Neurogenesis in the adult human hippocampus. Nat Med 4: 1313-1317

Golembiowska K, Dziubina A (2000) Effect of acute and chronic administration of citalopram on glutamate and aspartate release in the rat prefrontal cortex. Pol J Pharmacol 52: 441-448

Gould E, Tanapat P (1999) Stress and hippocampal neurogenesis. Biol Psychiatry 46: 1472-1479

Hellsten J, Wennstrom M, Mohapel P, Ekdahl CT, Bengzon J, Tingstrom A (2002) Electroconvulsive seizures increase hippocampal neurogenesis after chronic corticosterone treatment. Eur J Neurosci 16: 283-290

Honer WG, Falkai P, Bayer TA, Xie J, Hu L, Li HY, Arango V, Mann JJ, Dwork AJ, Trimble WS. (2002) Abnormalities of SNARE mechanism proteins in anterior frontal cortex in severe mental illness. Cerebral Cortex 12: 349-356

Malberg JE, Eisch AJ, Nestler EJ and Duman RS (2000) Chronic antidepressant treatment increases neurogenesis in adult rat hippocampus. J Neurosci 20: 9104-9110

Nilsson M, Perfilieva E, Johansson U, Orwar O, Eriksson PS (1999) Enriched environment increases neurogenesis in the adult rat dentate gyrus and improves spatial memory. J Neurobiol 39: 569-578

Stewart CA, Reid IC (2000) Repeated ECS and fluoxetine administration have equivalent effects on hippocampal synaptic plasticity. Psychopharmacol 148: 217-223

Vaidya VA, Siuciak JA, Du F, Duman RS (1999) Hippocampal mossy fiber sprouting induced by chronic electroconvulsive seizures. Neurosci 89: 157-166

van Praag H, Christie BR, Sejnowski TJ, Gage FH (1999) Running enhances neurogenesis, learning, and long-term potentiation in mice. Proc Natl Acad Sci USA 96: 13427-13431

Yamada M, Higuchi T (2002) Functional genomics and depression research. Eur Neuropsychopharmacol 12: 235-244

Yamada M, Takahashi K, Tsunoda M, Nishioka G, Kudo K, Ohata H, Kamijima K, Higuchi T, Momose K, Yamada M (2002) Differential expression of VAMP2/synaptobrevin-2 after antidepressant and electroconvulsive treatment in rat frontal cortex. Pharmacogenomics J 2: 377-382

Yamada M, Yamada M, Yamazaki S, Nara K, Kiuchi Y, Ozawa H, Yamada S,

Oguchi K, Kamijima K, Higuchi T, Momose K (2001) Induction of cysteine string protein after chronic antidepressant treatment revealed by ADRG microarray. Neurosci Lett 301: 183-186

S-100B Expression in Neonatal Rat Cortical "Barrels" and Thalamic "Barreloids"

Katsumasa T. Muneoka[1], Hisayuki Funahashi[1, 3], Tetsuo Ogawa[1], Makiko Kuwagata[1], Patricia M. Whitaker-Azmitia[2], and Seiji Shioda[1]

[1]Department of Anatomy I, Showa University School of Medicine, 1-5-8 Hatanodai, Shinagawa-Ku, Tokyo 142-8555, Japan
[2]Department of Psychology, State University of New York at Stony Brook, Stony Brook, NY 11794-2500, USA
[3]Division of Endocrinology, Beth Israel Deaconess Medical Center, Harvard Medical School, Boston, MA 02215, USA

Summary. "Barreloids" and "barrels" are patch-like structure representing single whiskers observed in the ventroposterior thalamic nucleus (VP) and somatosensory cortex in rodents, respectively. They are characteristic structure observed during the early neonatal period, which is an important period for the development of the networks in VP and somatosensory cortex. Various neurotransmitter systems have been reported to be involved in the development of these structure including, serotonergic, glutamatargic, GABAergic and cholinergic ones. The present study indicated that immunoreactivity for a calcium binding protein, S-100B, was transiently found as "barreloids" in VP at postnatal day (PND) 7 and as "barrels" in the somatosensory layer IV at PND 15, respectively. "Barrel" - like 5-hydroxytriptamine transporter (5-HTT) immunoreactivity was also found in the somatosensory cortex at PND 15. Morphological findings indicated that S-100B was present in cellular nuclei and released into extracellular space in both the thalamus and cortex. S-100B is suggested to be involved in the formation or remodeling of networks in VP and the somatosensory cortex via modulating calcium levels intra- and extracellularly beside changes in related neurotransmitters.

Key words. S-100B, 5-hydroxytriptamine (5-HT), 5-HT transporter, Ventroposterior thalamus, Somatosensory cortex

1 "Barreloids" and "Barrels" during the Neonatal Period

The ventroposterior thalamic nucleus (VP) is a relay center transferring sensory information from the trigeminal nucleus to the somatosensory cortex. In rodents, patch-like structure is visualized with metabolic markers, cytochrome oxidase or succinic dehydrogenase in VP as thalamic "barreloids" (Haidarliu and Ahissar, 2001; Muñoz et al., 1999; Persico et al., 2001) and in the somatosensory cortex as cortical "barrels" (Riddle et al., 1992). Each "barreloid" and "barrel" represents single whiskers in VP and somatosensory cortex, respectively. In the first week of age in rodents, VP is morphologically immature (Matthews et al., 1977). Various morphological as well as functional development occurs during and after this period (Agmon et al., 1995; Matthews et al., 1977; Scheibel et al., 1976; Zantua et al., 1996). In addition, various neurotransmitter systems are involved in the development of "barreloids" or "barrels" during the early neonatal period such as the serotonergic (Persico et al., 2001; Xu et al., 2004), glutamatergic (Muñoz et al., 1999), GABAergic (Frassoni et al., 1991), cholinergic ones (Broide et al., 1995).

2 S-100B and 5-HT Transporter (5-HTT) in Neonatal Brains

2.1 Methods

At postnatal day (PND) 7, 15 or 21, male Sprague-Dawley rats were anesthetized and perfused with saline followed by 4 % periadate-lysin-paraformaldehyde. Brains were postfixed, immersed in 20 % sucrose and sectioned on a sliding microtome with a frozen stage. Immunoreactions were performed with the free-floating methods. Sections were incubated for 72 hours at 4° C with a monoclonal anti - S-100B (1:1000, SIGMA) or polyclonal anti - 5-HTT antibody (1:1000, gifted from Prof. Zhou, Indiana University) in phosphate buffered saline containing 0.3 % Triton X-100 and 0.3 % normal goat serum. After washing, sections were incubated for 2 hours with biotinylated mouse IgG or biotynylated rabbit IgG at room temperature. The sections were washed visualized with the Vector Elite ABC kit with 0.05% diaminobenzidine.

2.2 Results

A segmented S-100B immunostaining was observed as "barreloids" in VP at PND 7 (Fig. 1A) whereas S-100B staining in VP became more homogeneous at PND 15 and later. In the somatosensory cortex, "barrel" -like staining for S-100B was observed in layer IV at PND 15 (Fig. 1B) but not at PND 7 or 21. In higher magnification in VP (Fig. 1A, inlet) and the cortex (Fig. 1B, inlet), intense staining in cellular nuclei and diffuse staining surrounding them suggesting a release of S-100B into the extracellular space were observed. Although 5-HTT immunoproduct was not detected in VP at PND 7 (Fig. 1C) and 15, intense staining was detected in the internal capsule (ic) (Fig. 1C). Patch - like staining for 5-HTT was found in the layer IV and IV at the somatosensory cortex at PND 7 indicating 5-HTT - positive "barrels" (Fig. 1D) whereas no patch - like structure was detected at PND 15 and later.

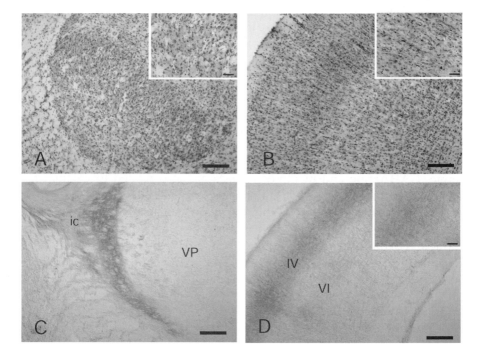

Fig. 1. S-100B staining counterstained with cresyl violet in ventroposterior thalamic nucleus (*VP*) and postnatal day (PND) 7 (**A**) and in the somatosensory cortex at PND 15 (**B**).5HTT staining in VP (**C**) and the somatosensory cortex (**D**) at PND 7. *Scale bars* = 200 μm and 50 μm (inlets). *ic*, internal capsule; *IV*, layer IV; *VI*, layer VI

3 Possible role of S-100B in the Developing VP and Somatosensory Cortex

The present study indicated that S-100B protein was shown in VP and the somatosensory cortex layer IV as "barreloids" and "barrels", respectively. Data indicate that S-100B is present in cellular nuclei and is released into the extracellular space as suggested (Muneoka et al., 2003). Results also showed "barrels" - like expression of 5-HTT in the somatosensory cortex as reported (Persico et al., 2001; Xu et al., 2004). Expression of 5-HTT accompanied with reuptake of 5-HT (Lebrand et al., 1993; Xu et al., 2004) and of 5-HT$_{1B}$ receptor (Young-Davies et al., 2000) on axons and terminals of thalamocortical neurons during the neonatal period has been reported. 5-HT, 5-HTT and 5-HT$_{1B}$ receptor are involved in the formation and refinement of the barrel field (Persico et al., 2001; Rebsam et al., 2002; Xu et al., 2004; Young-Davies et al., 2000). S-100B is a calcium binding protein and released from astrocytes via stimulation of 5-HT$_{1A}$ receptor (Whitaker-Azmitia, 2001). Released S-100B into extracellular space could modulate intra and extracellular calcium levels in glia and neurons (Barger and VanEldik, 1992) and influence glutamatergic signals (Ciccarelli et al., 1999). Recent work demonstrated a spontaneous change in intracellular calcium and their propagation to neighboring cells as a "calcium wave" in developing glia and neurons (Parpura and Haydon, 2000). The calcium wave was observed in the neonatal thalamus accompanied with glutamate release (Parri and Crunelli, 2002). Furthermore, neurotransmitter systems relating to the development of "barreloids" or "barrels" that are glutamatergic, GABAergic and nicotinergic ones are also related to calcium dynamics (Berninger et al., 1995). Taken together, S-100B expressed and released in neonatal "barreloids" and "barrels" is suggested to be involved in the formation or remodeling of thalamocortical networks via modulating intra- and extracellular calcium levels beside developmental changes in related neurotransmitters.

References

Agmon A, Yang LT, Jones EG, O'Dowd DK (1995) Topological precision in the thalamic projection to neonatal mouse barrel cortex. J Neurosci 15: 549-561

Barger SW, VanEldik LJ (1992) S100ß stimulates calcium fluxes in glial and neuronal cells. J Biol Chem 267: 9689-9694

Berninger B, Marty S, Zafra F, Berzaghi MdP, Thoenen H, Lindholm D (1995) GABAergic stimulation switches from enhancing to repressing BDNF expression in rat hippocampal neurons during maturation in vitro. Development 121:

2327-2335

Broide RS, Robertson RT, Leslie FM (1996) Regulation of alpha$_7$ nicotinic ace-tylcholine receptors in the developing rat somatosensory cortex by thalamo-cortical afferents. J Neurosci 16: 2956-2971

Ciccarelli R, DiIorio P, Bruno V, Battaglia G, D'alimonte I, D'onofrio M, Nicoletti F, Caciagli F (1999) Activation of A$_1$ adenosine or mGlu3 metabotropic glu-tamate receptors enhances the release of nerve growth factor and S-100ß pro-tein from cultured astrocytes. Glia 27: 275-281

Frassoni C, Bentivoglio M, Spreafico R, Sánchez MP, Puelles L, Fairen A (1991) Postnatal development of calbindin and parvalbumin immunoreactivity in the thalamus of the rat. Dev Brain Res 58: 243-249

Haidarliu S, Ahissar E (2001) Size gradients of barreloids in the rat thalamus. J Comp Neurol 429: 373-387

Lebrand C, Cases O, Adelbrecht C, Doye A, Alvarez C, ElMestikawy S, Seif I, Gaspar P (1993) Transient uptake and storage of serotonin in developing tha-lamic neurons. Neuron 17: 823-835

Matthews MA, Narayanan CH, Narayanan Y, Onge MFS (1977) Neuronal matura-tion and synaptogenesis in the rat ventrobasal complex: alignment with devel-opmental changes in rate and severity of axon reaction. J Comp Neurol 173: 745-772

Muneoka KT, Borella A, Whitaker-Azmitia PM (2003) Transient expression of S-100ß immunostaining in developing thalamus and somatosensory cortex of rat. Dev Brain Res 142: 101-104

Munõz A, Liu X-B, Jones EG (1999) Development of metabotropic glutamate re-ceptors from trigeminal nuclei to barrel cortex in postnatal mouse. J Comp Neurol 409: 549-566

Parpura V, Haydon PG (2000) Physiological astrocytic calcium levels stimulate glutamate release to modulate adjacent neurons. Proc Natl Acad Sci USA 97: 8629-8634

Parri HR, Crunelli V (2002) Astrocytes, spontaneity, and the developing thalamus. J Physiol (Paris) 96: 221-230

Persico AM, Menguel E, Moessner R, Hall SF, Revay RS, Sora I, Arellano J, De-Felipe J, Giménez-Amaya J, Conciatori M, Marino R, Baldi A, Cabib S, Pas-cucci T, Uhl GR, Murphy DL, Lesch KP, Keller F (2001) Barrel pattern for-mation requires serotonin uptake by thalamocortical afferents, and not vesicular monoamine release. J Neurosci 21: 6862-6873

Rebsam A, Seif I, Gaspar P (2002) Refinement of thalamocortical arbors and emergence of barrel domains in the primary somatosensory cortex: a study of normal and monoamine oxidase a knock-out mice. J Neurosci 22: 8541-8552

Riddle D, Richards A, Zsuppan F, Purves D (1992) Growth of the rat somatic sen-sory cortex and its constituent parts during postnatal development. J Neurosci 12: 3509-3524

Scheibel ME, Davies TL, Scheibel AB (1976) Ontogenetic development of soma-tosensory thalamus I. Morphogenesis. Exp Neurol 51: 392-406

Whitaker-Azmitia PM (2001) Serotonin and brain development: Role in human developmental diseases. Brain Res Bull 56: 479-485

Xu Y, Sari Y, Zhou FC (2004) Selective serotonin reuptake inhibitor disrupts organization of thalamocortical somatosensory barrels during development. Dev Brain Res 150: 151-161

Young-Davies CL, Bennett-Clarke CA, Lane RD, Rhoades RW (2000) Selective facilitation of the serotonin(1B) receptor causes disorganization of thalamic afferents and barrels in somatosensory cortex of rat. J Comp Neurol 425: 130-138

Zantua JB, Wasserstrom SP, Arends JJ, Jacquin MF, Woolsey TA (1996) Postnatal development of mouse "whisker" thalamus: ventroposterior medial nucleus (VPM), barreloids, and their thalamocortical relay neurons. Somatosens Mot Res 13: 307-322

Postnatal Change of Glycinergic Synaptic Transmission from Supratrigeminal Region to Trigeminal Motoneurons

Tomio Inoue[1], Shiro Nakamura[1], Kan Nakajima[2], and Kotaro Maki[2]

Departments of [1]Oral Physiology and [2]Orthodontics, Showa University School of Dentistry, 1-5-8 Hatanodai, Shinagawa-ku, Tokyo 142-8555, Japan

Summary. We have investigated excitatory synaptic transmission from the lateral part of the supratrigeminal region (lSuV) to the trigeminal motor nucleus (MoV) in neonatal and juvenile rat brain stem slice preparations by high-speed optical recording techniques and gramicidin perforated-patch recordings. Electrical stimulation of lSuV evoked optical responses in MoV. An antidromic response in lSuV was evoked by MoV stimulation while synaptic transmission was suppressed by substitution of external Ca^{2+} with Mn^{2+}. Application of CNQX and APV to MoV reduced the optical responses in MoV evoked by lSuV stimulation in both neonatal and juvenile rats. Application of strychnine to MoV also suppressed the optical responses in MoV of neonatal rats. On the other hand, strychnine enhanced the optical responses in MoV of juveniles. Gramicidin perforated-patch recordings from trigeminal motoneurons (TMNs) revealed that glycinergic postsynaptic potentials evoked by lSuV stimulation were depolarizing in neonatal rats but become hyperpolarizing in juveniles. We conclude that inputs from lSuV excite TMNs through activation of glutamate or glycine receptors in neonatal rats, whereas glycine receptor activation in TMNs becomes inhibitory in juveniles. Such postnatal change of synaptic transmission from lSuV to MoV might be involved in the transition from suckling to mastication.

Key words. Supratrigeminal region, optical recording, gramicidin, development, feeding behavior

1 Introduction

Feeding behavior in mammals changes from suckling to mastication during development and the neuronal circuits controlling feeding behavior should change in parallel to the development of oro-facial structures. In the present study, we have investigated the postnatal change in the neuronal circuit of rat brainstem slices which is involved in increasing jaw muscle activities, by high-speed optical recording and gramicidine perforated-patch recording techniques.

2 Materials and methods

Transverse brainstem slices (500 μm) including the trigeminal motor nucleus (MoV) were obtained from neonatal and juvenile Wistar rats (P1–12). Slices were stained for 1 hr with the voltage-sensitive dye Di-4-ANEPPS (100 μg/ml in ACSF). Optical signals were recorded by a CCD camera system (MiCAM01, Brain Vision) with an acquisition time of 3.0 ms and averaged over 128 trials. A tungsten electrode was used to deliver extracellular stimuli (200 μs; 20–30 μA) in the brainstem slices. CNQX, APV, bicuculline and strychnine were focally applied to MoV by using an 8-channel perfusion system (BPS-8, ALA). The composition of ACSF was (in mM) 130 NaCl, 3 KCl, 2 $MgCl_2$, 1.25 NaH_2PO_4, 26 $NaHCO_3$, 2 $CaCl_2$, and 10 D-glucose.

Gramicidin perforated-patch clamp recordings were made from TMNs in 300–500 μm transverse slice preparation (P1–P10) with an Axoclamp 200B amplifier (Axon Instruments). The patch pipette contained 150 KCl, 10 HEPES, and 5 QX314. Gramicidin was dissolved in DMSO (10 mg/ml) and was diluted into the pipette solution at a final concentration of 20 μg/ml just before use.

3 Results

In order to find the region sending excitatory outputs to MoV, we systematically applied a single pulse stimulation to various sites on the slices and observed whether the optical response was evoked in MoV of P1–P6 rats. Stimulation of the lateral part of the supratrigeminal region (lSuV) evoked optical responses in MoV at mean (± SD) latencies of 10 ± 5.3 ms ($n = 16$), whereas stimulation of sites other than lSuV did not evoke optical responses in MoV. Then we stimulated MoV while synaptic transmission was suppressed by substitution of external Ca^{2+} with Mn^{2+}. An antidromic optical response in lSuV was observed in the time frame next to the frame in which the stimulation was delivered, and it disappeared in the following frame, suggesting that the neurons located in lSuV send their axons to MoV. To investigate which neurotransmitters are involved in the synaptic transmission from lSuV to MoV, we examined the effects of amino acid receptor antagonists on the optical responses in MoV evoked by lSuV stimulation. Focal application of CNQX (20 μM) combined with APV (20 μM) to MoV for 20 min reduced the optical responses in MoV by 43 ± 21% ($n = 17$, $P < 0.01$) in P1–P6 rats (Fig.1 *left*). Interestingly, focal application of strychnine (5–20 μM) also reduced the optical responses in MoV by 55 ± 20% ($n = 16$, $P < 0.01$). The optical responses in MoV recovered to more than 83% of the control level 20 min after the washout of antagonists. Application of bicuculline (10-20 μM) did not alter the optical responses in MoV. These results suggest that both glutamatergic and glycinergic inputs from lSuV to TMNs are likely excitatory in P1–P6 rats.

Next, we examined whether effects of those inputs on TMNs change during development. In P7–P12 rats, lSuV stimulation also evoked optical responses in MoV, and application of CNQX and APV to MoV reduced the optical responses by 57 ± 13% ($n = 10$, $P < 0.01$, Fig.1 *right*). The rates of suppression in the optical responses after antagonists application were not different between P1–P6 rats and P7–P12 rats ($P > 0.2$). On the other hand, application of strychnine to MoV enhanced the optical responses by 19 ± 23% ($n = 14$, $P < 0.01$) in P7–P12 rats. The rate of change in the optical responses of P7–P12 rats was significantly different from that of P1–P6 rats ($P < 0.01$, compare Fig.1 *left* and *right*). These results suggest that glycinergic inputs to TMNs appeared to be inhibitory in P7–P12 rats.

Then we confirmed that glycinergic postsynaptic potentials (PSPs) in TMNs induced by glycinergic inputs from lSuV change from excitatory to inhibitory during postnatal development by using gramicidin perforated-patch recording techniques. In the presence of 20 μM CNQX and 20 μM APV, depolarizing PSPs were recorded in TMNs at the membrane potential of –60 mV in P1–P3 rats. The PSPs were abolished by bath application of 10 μM strychnine. Postsynaptic currents in TMNs evoked by lSuV stimulation were recorded at various holding potentials and had a reversal potential of –31 ± 8 mV ($n = 5$) in P1–P3 rats. On the other hand, the postsynaptic currents had a reversal potential of –76 ± 6 mV ($n = 4$) in P9–P10 rats which is lower than the resting membrane potential in adult TMNs (–67.3 ± 0.6 mV, Inoue et al. 1999). These results suggest that glycinergic inputs to TMNs from lSuV were excitatory in P1–P3 rats but changed to inhibitory in P9–P10 rats.

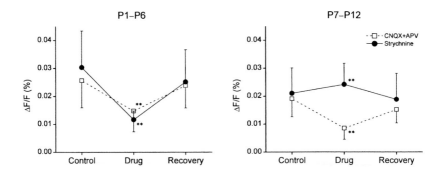

Fig. 1. Postnatal changes in effects of amino acid receptor antagonists on optical responses in MoV evoked by lSuV stimulation. Application of CNQX and APV (*open squares*) reduced optical responses in MoV of both P1–P6 and P7–P12 rats, whereas strychnine (*filled circles*) enhanced optical responses in P7–P12 rats. ** P < 0.01 *vs* control

4 Discussion

It has been suggested in anesthetized adult cats that the supratrigeminal region contains glycinergic inhibitory last-order interneurons for jaw-closing motoneurons (Goldberg and Nakamura 1968, Kidokoro et al. 1968, Nakamura et al. 1973). The present results in P7–P12 rats are in accord with the results of those studies, whereas glycinergic inputs from lSuV were excitatory in P1–P6 rats. It has been shown that application of GABA or glycine depolarizes neonatal motoneurons of the spinal cord (Fulton et al. 1980) and brainstem (Singer et al. 1998). Such GABAergic or glycinergic depolarization is most likely caused by Cl⁻ efflux due to the high intracellular Cl⁻ concentration maintained in immature neurons (see Owens et al. 1996, Ben-Ari 2002). Such postnatal change of synaptic transmission from lSuV to MoV might be involved in the transition from suckling to mastication.

5 Acknowledgements

This study was supported in part by a Showa University Grant-in-Aid for Innovative Collaborative Research Projects and a Special Research Grant-in-Aid for Development of Characteristic Education from the Japanese

Ministry of Education, Culture, Sports Science and Technology.

References

Ben-Ari Y (2002) Excitatory actions of gaba during development: the nature of the nurture. Nat Rev Neurosci 3: 728-739

Fulton BP, Miledi R, Takahashi T (1980) Electrical synapses between motoneurons in the spinal cord of the newborn rat. Proc R Soc Lond B Biol Sci 208: 115-120

Goldberg LJ, Nakamura Y (1968) Lingually induced inhibition of masseteric motoneurones. Experientia 24: 371-373

Inoue T, Itoh S, Kobayashi M, Kang Y, Matsuo R, Wakisaka S, Morimoto T (1999) Serotonergic modulation of the hyperpolarizing spike afterpotential in rat jaw-closing motoneurons by PKA and PKC. Journal of Neurophysiology 82: 626-637

Kidokoro Y, Kubota K, Shuto S, Sumino R (1968) Possible interneurons responsible for reflex inhibition of motoneurons of jaw-closing muscles from the inferior dental nerve. J Neurophysiol 31: 709-716

Nakamura Y, Mori S, Nagashima H (1973) Origin and central pathways of crossed inhibitory effects of afferents from the masseteric muscle on the masseteric motoneuron of the cat. Brain Res 57: 29-42

Owens DF, Boyce LH, Davis MB, Kriegstein AR (1996) Excitatory GABA responses in embryonic and neonatal cortical slices demonstrated by gramicidin perforated-patch recordings and calcium imaging. J Neurosci 16: 6414-6423

Singer JH, Talley EM, Bayliss DA, Berger AJ (1998) Development of glycinergic synaptic transmission to rat brain stem motoneurons. J Neurophysiol 80: 2608-2620

Central Nervous System Concerned with the Stress Induced Inhibition of Cellular Immune Activity

Takao Sato, Shiyu Guo, and Tadashi Hisamitsu

Department of Physiology, School of Medicine, Showa University, 1-5-8 Hatanodai, Shinagawa-ku, Tokyo 142-8555, Japan

Summary. Many studies about the effect of stressors on immune activities were reported, but the mechanisms of those effects were obscure yet. Restraint water immersion stress reduced splenic NK cell activity (NK activity) measured by cytotoxicity to YAC-1 cells in rats through both humoral and neural mechanisms. Electrical lesion of hypothalamic paraventricular nucleus (PVN) markedly enhanced NK activity, but the lesion of ventromedial hypothalamic nucleus (VMH) depressed NK activity. These results suggest that the stress may inhibit NK activity through the activation of hypothalamic paraventricular nucleus.

Key words. Restraint water immersion stress, NK cell activity, splenic sympathetic nerve, PVN, VMH

1 Introduction

Many studies about the effect of stressors on immune activities were reported, but the mechanisms of those effects were obscure yet. In this study we attempted to clarify the central mechanism by which stressor might inhibit cellular immune activity in the rats. As the candidate we chose hypothalamic paraventricular nucleus in which we recognized c-fos expression by LPS stress.

2 Methods

Restraint water immersion stress (stress) for 20 hours was used as stressor. The stress was established as a sure method causing digestive ulcer in rats.

The details were described in our previous report. Cellular immune activity was assessed by cytotoxicity (% Specific Lysis: NK activity) of splenic NK cell (target cell: T) to YAC-1 cell (effector cell: E). The cytotoxicity was measured by release of chromium 51 from YAC-1 cell. Denervation of splenic sympathetic nerve (Denervation) was manipulated by removing sympathetic nerve around the splenic artery under pentobarbital anesthesia (50mg/kg, ip). Electrical lesion of PVN was managed by passing anodal current (0.1mA, 30sec) under pentobarbital anesthesia. One week after the lesion for recovering time, the spleen was isolated under pentobarbital anesthesia (50mg/kg, ip). The location of central lesion was confirmed by cryostat section. The statistical significance was confirmed by 2-way ANOVA.

3 Results

Stress remarkably reduced NK activity with statistical significance (p=0.0001). NK activities (mean+-SE) of control animals were 20.0+-4.4, 30.3 ± 9.1 and 36.9 ± 8.0 at 25:1, 50:1 and 100:1 of E:T ratio respectively. Those of stressed animals were 5.9 ± 2.4, 8.8 ± 2.4 and 11.9 ± 3.7. The serum collected from stressed rats reduced NK activity of other non-stressed rats with statistical significance (p=0.0001). NK activity at 100:1 of E:T ratio in the serum from non-stressed rat was 59.3 ± 2.4. NK activity in the serum from stressed rat was 45.9 ± 3.0. Denervation partially restored the reduced NK activity by stress with statistical significance (p = 0.0032). NK activities of intact stressed animals were 13.1±0.7, 15.6±1.4 and 23.7±1.2 at 25:1, 50:1 and 100:1 of E:T ratio respectively. Those of denervated stressed animals were 18.5 ± 3.8, 28.4 ± 4.1 and 35.3 ± 4.7 respectively.

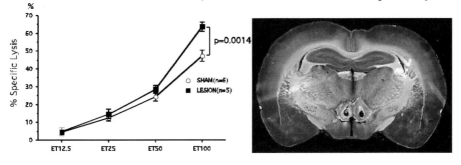

Fig. 1. Enhancement of NK cytotoxicity by PVN lesion in rats

Fig. 2. Location of PVN lesion

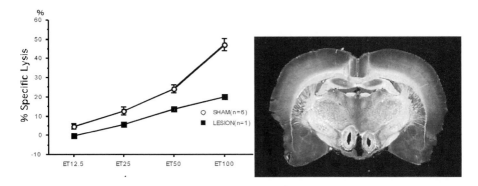

Fig. 3. Attenuation of NK cytotoxicity by VMH lesion in rats

Fig. 4. Location of VMH lesion

Lesion of PVN enhanced NK activity (Fig.1). The cryostat section of PVN lesion was shown in Fig.2. Lesion of hypothalamic ventromedian nucleus (VMH) reduced NK activity (Fig.3). The cryostat section of VMH lesion was shown in Fig.4.

4 Discussion

The result that the serum collected from stressed rats reduced NK activity of other non-stressed rats indicates that stress inhibited NK activity through humoral mechanism. The result that denervation restored the reduced NK activity by stress means that stress diminished NK activity through sympathetic nervous system. It is shown that the central nervous system concerned with inhibition of NK activity by stress might be PVN because PVN lesion enhanced NK activity. On the other hand, VMH might accelerate NK activity.

These results suggest that restraint water immersion stress reduced NK activity through both neural and humoral mechanism, central inhibiting site may exist in PVN, and central accelerating site may exist in VMH.

References

Seiber WJ, Rodin J, Larson L, et al. (1992) Modulation of human natural killer cell activity by exposure to uncontrollable stress. Brain Behav Immun 6: 141-156

Goodkin K, Fuchs I, Feaster D, et al. (1992) Life stressors and coping style are associated with immune measures in HIV-1 infection--a preliminary report. Int J

Psychiatry Med 22: 155-172

Locke SE, Kraus L, Leserman J, et al. (1984) Life change stress, psychiatric symptoms, and natural killer cell activity. Psychosom Med 46: 441-453

Frawzy FI, Frawzy NW, Hyun CS et al. (1993) Malignant melanoma. Effect of an early structured psychiatric intervention, coping, and affective state on recurrence and survival 6 years later. Arch Gen Psychiatry 50: 681-689

Persky VW, Kempthorne-Rawson J, Shekelle RB (1987) Personality and risk of cancer: 20-year follow-up of the Western Electric Study. Psychosom Med 49: 435-449

Irwin M, Daniels M, Smith T et al. (1987) Impaired natural killer cell activity during bereavement. Brain Behav Immun 1: 98-104

Bartrop RW, Luckhurst E, Lazarus L et al. (1977) Depressed lymphocyte function after bereavement. Lancet 1: 834-836

Zisook S, Shuchter SR, Irwin M et al. (1994) Bereavement, depression, and immune function. Psychiatry Res 52: 1-10

Xu S, Guo S, Jiang X et al. (2003) Effect of indomethacin on the c-fos expression in AVP nand TH neurons in rat brain induced by lipopolysaccharide. Brain Res. 966: 13-18

Sato T, Jiao YY, Kasahara T et al. (1999) Restraint water immersion stress suppresses the natural killer cytotoxicity of the spleen in rats. Jap J Stress Sci. 14: 67-76

Sato T, Yu Y, Guo SY et al. (1996) Acupuncture stimulation enhances splenic natural killer cytotoxicity in rats. Jpn J Physiol 46: 131-136

Part IV
Brain Functions by the Dipole Tracing Method

Can not Live without Breathing, without Emotions

Yuri Masaoka and Ikuo Homma

Department of Physiology, Showa University School of Medicine, 1-5-8 Hatanodai, Shinagawaku, Tokyo 142-8555, Japan

Summary. Respiration is spontaneous activity regulated in the brainstem for metabolic purposes. However, respiratory activity can be changed voluntarily, which could be related to supplementary and pre-motor areas in the cortex (Ramsay et al. 1993). In addition, breathing is altered unconsciously by many types of cognitive inputs such as auditory and visual stimuli (Mador and Tobin 1991). Not only these external stimuli, but also emotions or the internal state changes breathing patterns. Therefore, the final respiratory outputs could be results produced from an interaction between metabolism and higher functions, namely, the brainstem and the limbic and cortical structures. Respiratory psychophysiology studies in humans have reported that various emotions alter respiratory patterns (Boiten et al. 1994). On the other hand, recent neuroimaging studies have investigated the neuroanatomical correlates of emotions, especially fear and anxiety (Morris et al. 1998; Reiman et al. 1989). Emotions are not phenomena within the brain itself but involve physiological changes of the whole body. Respiration is one of the physiological outputs altered by emotions.
Besides metabolic, emotional and volitional breathing, respiration has another function, the role of olfaction. Smelling is largely dependent on inhaling; we are able to sense an odor by sniffing. In this chapter we focus on the relation between the limbic areas and respiration by showing our results obtained in humans. In addition, we also report results from our previous study regarding new insights into respiration through olfaction.

Key words. Respiration, anxiety, limbic area, olfaction

1 Amygdala: A role of Emotion and Breathing Frequency

A human study has reported that the amygdala plays a role in emotional processing, especially negative emotions such as fear and anxiety (Davis 1992). In an animal study, electrical stimulation of the amygdala elicited specific signs of fear which included various autonomic and physiological responses. In respiratory physiology Harper et al. (1984) reported that electrical stimulation of the amygdala increases respiratory frequency. Investigation between deep structures, such as the amygdala, and respiration in humans is limited; however, we report studies which show the relationship between the amygdala and respiration in normal subjects and epilepsy patients with lesions of the left amygdala.

1.1 Effect of Anticipatory Anxiety on Breathing and their Related Areas in the Brain

In a laboratory situation anticipatory anxiety was produced in subjects while measuring minute ventilation, tidal volume, respiratory rate, end-tidal CO_2 and O_2 consumption (Masaoka and Homma, 2001). The subjects were informed that electrical stimulation would be delivered within two minutes after the onset of a warning light. "Anticipatory anxiety" was defined as the time between warning onset and stimulation. During anticipatory anxiety, respiratory rate increased without changes in O_2 consumption,

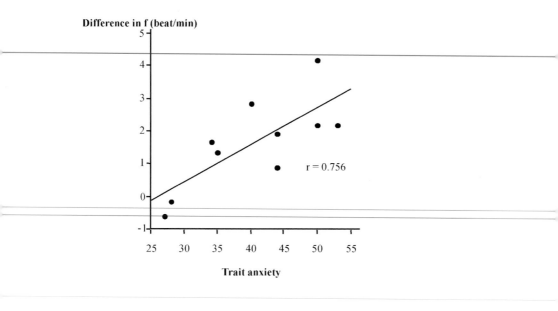

Fig. 1. Correlation between breathing frequency (f) and trait anxiety scores (Masaoka and Homma, 2001)

and there was a positive correlation between respiratory rate and individual trait anxiety scores (Fig. 1). The results indicate that increasing respiratory rate is not caused by metabolic changes, and the level of increasing rate depends on individual trait anxiety. Anxiety, which could be associated with the limbic area, dominantly affects respiratory outputs.

The dipole tracing method which incorporates a scalp-skull-brain head model, estimated the dipole location from the surface electrodes on the scalp (Fig. 2) (Homa et al. 1994; Homma et al. 2001). The onset of inspirations during anticipatory anxiety were triggered for averaging EEG, and the areas related to anxiety and respiratory rate were investigated. From 300 to 350 ms after the inspiratory onsets, positive waves, referred to as Respiratory-related Anxiety potential (RAP), were observed. SSB/DT estimated dipoles in the right temporal pole, the right temporal pole and the left amygdala in the most anxious subjects.

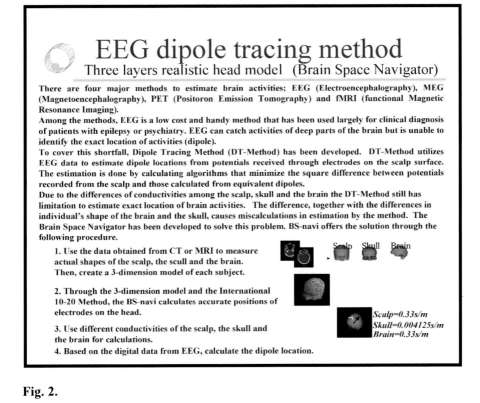

Fig. 2.

1.2 Anticipatory anxiety and physiological responses in patients with the left amygdala lesion

Anticipatory anxiety was also tested in two patients with epilepsy whose diagnosis showed the foci of epileptic spikes located in the left amygdala. We sought direct evidence of the effect of anticipatory anxiety on respiration, perspiration and heart rate before and after lesion surgery of the left amygdala. A lesion in the left amygdala resulted in decreases of trait and state anxiety, respiratory rate, and activity in the right amygdala in both subjects; one subject also showed notable decreases in skin conductance and heart rate (Fig. 3). The study also shows that activities in the right amygdala before the lesion were not observed after the lesion. This study suggests that activity of the right amygdala is dominantly activated in anxiety and anxiety-related physiological responses but needs excitatory inputs from the left amygdala.

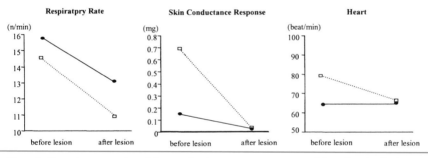

Fig. 3. Respiratory rate, skin conductance responses, and heart rate during anticipatory anxiety before and after lesion surgery of the left amygdala (Masaoka et al. 2003)

3 What is Anxiety?

There is a well-known debate between the James-Lang and the Cannon-Bird theories of emotion which question whether feelings cause bodily responses or bodily responses cause feelings. Does anxiety enhance the respiratory rate or does respiratory rate produce anxiety? From our data, increasing respiratory rate during the situation of emergency (state of expecting electrical stimuli) occurs unconsciously, and this change involves activation of the temporal pole and the amygdala. Electrical stimulation on the left amygdala, in a patient who had deep electrodes installed to evaluate the location of the epileptic spike, showed an increase of respiratory frequency (Masaoka et al. 2004). Stimulation of the deep electrodes was performed in the course of a routine examination by asking the patient whether or not the stimulation caused a feeling similar to that before the epileptic seizure. The patient reported that she felt an uncertain unpleasantness which she could not explain, and the border between external objects and herself became indistinguishable.

External stimuli reached the amygdala by proceeding a pathway from the thalamus to the sensory cortex to the amygdala. LeDoux (1998) shows studies of conditioned emotional responses of fear in which unconscious evaluation of external stimuli bypass the sensory cortex and follow a direct pathway from the thalamus to the amygdala threby responding to dangerous external stimuli more quickly.

Increasing respiratory rate could be caused by unconscious evaluation in the amygdala (ex., umbiguity, or an uncertain external signal), and the emotion of "anxiety" is the state before conscious representation or emotional meaning of external events. In other words, anxiety is the state in which we try to interpret or label an event which caused physiological responses. If we become conscious of meaning a physiological response or we can express the emotional feeling, maybe we are no more in the category of "anxiety." In this sense, anxiety could be the emotion which accompanies physiological responses, and the state of emotion parallels these physiological responses.

Which area is related to emotional consciousness or awareness?

3 Respiration in Olfaction

Olfactory information ascends directly to the limbic areas and is not relayed in the thalamus. Olfactory cells have cilia (dendrites) that extend from the cell body into the nasal mucosa. The axons carry impulses to the olfactory bulb when the olfactory receptor is stimulated. The olfactory bulb sends signals to the prepiriform and the piriform cortex, which are called the primary olfactory cortex, anterior olfactory nucleus, amygdaloid nucleus, olfactory tubercle and entorhinal cortex. Perception of odors is dependent on respiration; our sense of smell is enhanced by inhalation or inspiration. Figure 4 shows respiratory maneuvers during pleasant and unpleasant odor stimuli.

A pleasant odor increases tidal volume and decreases respiratory rate; on the other hand, an unpleasant odor decreases tidal volume and increases respiratory rate. These changes were elicited unconsciously and were not caused by metabolism. Odor information is projected directly to the olfactory and the limbic areas, and this stimulation alters respiratory patterns.

Unpleasant

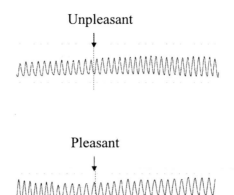

Pleasant

Fig. 4. Effects of unpleasant and pleasant odor on repiratory patterns

3.1 Inspiration phase-locked alpha band oscillation and their dipole locations

Event-related potentials are identified from changes in the EEG that are related to the occurrence of external stimuli, such as somato-sensory, auditory or visual signals. If the inspiratory onset during odor stimuli is used as a trigger for averaging EEG, we could determine the inspiration-related potential related to olfaction. In addition, we can estimate the dipole location from the averaged potentials by means of SSB/DT (Masaoka et al. 2005).

Individual detection and recognition levels of pleasant and unpleasant odors were tested. Each odorant was dissolved in propylene glycol at eight different concentrations; the trials began with the lowest concentration and were repeated with progressively higher concentrations. The concentration at which the odor was perceived but not identified was considered the "detection level." As the concentration increased, the subject was able to identify the odor. The subject was required to identify each odor and describe the kind of odor. The concentration at which an odor was first identified was considered the "recognition level."

Figure 5 shows a typical example of the averaged potentials which triggered the inspiration onset during odor stimuli. In the expiration phase, 3-

4 negative and positive waves were not seen; however, in the inspiration phase, we can clearly see 3-4 waves.

Power spectra analysis indicates that these waves were categorized in the alpha band frequency and were phase-locked to inspiration. These waves are referred to as "Inspiration phase-locked alpha band oscillation" (I-α). I-αwas observed in both pleasant and unpleasant odors, and in both their threshold and recognition levels. The advantage of measuring EEG is that it has a temporal resolution on the order of a millisecond, and event-related potentials detect the latency of negative or positive waveforms related to the stimulus event. SSB/DT estimated the source generators of I-α in the entorhinal cortex; the posterior orbitofrontal cortex subserves odor detection, and the entorhinal cortex, hippocampus and anteorior orbitofrontal cortex subserve odor recognition (Fig. 6).

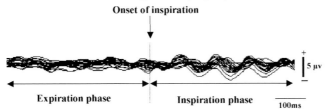

Fig. 5. Inspiratory phase-locked alpha oscillation (I-α) was observed after the onset of inspiration, and I-α was not observed in the expiratory phase (Masaoka et al. 2005)

The amygdala activates all odor stimuli; however, the amygdala is more active during unpleasant odor stimuli. Odor detection is a level categorized by detectable but unidentifiable smell; subjects cannot discern whether a smell is pleasant or unpleasant. Even at detection levels of unpleasant odors, respiratory rate increases. It is likely that respiration outputs respond more rapidly than cognition. Olfactory signals project directly to the amygdala, the entorhinal and the hippocampus immediately changing the breathing pattern, even before reaching the cognition level. At a recognition level, the orbitofrontal cortex is active for both pleasant and unpleasant odors. Several fMRI studies report that the orbitofrontal cortex is consistently activated by odor. The link between the entorhinal and orbitofrontal cortex could play a role in the representation of emotion or give an emotional meaning to odor. Therefore, recognition of an odor is not an emotion of anxiety but could be categorized as another emotion such as pleasantness or unpleasantness.

Unpleasant odor

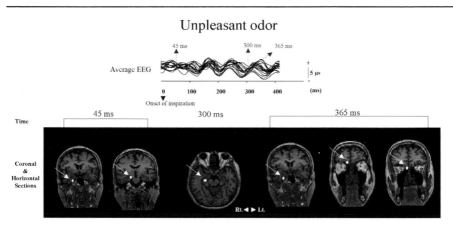

Fig. 6. Example of dipole locations estimated during unpleasant odor-recognition superimposed on a subject's coronal and horizontal MRI sections. Rt, right; Lt, left for the coronal and horizontal sections. Dipole location: at 45 ms, RT entorhinal cortex and RT hippocampus; 300 ms, RT amygdala; 365 ms, RT entorhinal cortex, and orbitofrontal cortex (Masaoka et al. 2005)

3.2 Olfactory-Related Areas and Cortical Oscilallation

I-α originates from the olfactory-related areas in this study. Alpha rhythms are generated from interaction with the thalamocortical and cortico-cortical systems. Brain rhythms consist of several oscillation types generated in interacting cortical and thalamic neuronal networks, and synchronization of low frequency cortical oscillations reflects a behavioral state associated with a brain disconnected from the external environment (Steriade, 2000). However, this could not be the case in the olfactory system. Fontanini et al. (2004) examined slow oscillations in the olfactory system in ketamine-xylazine anesthetized rats and reported a strong relation between the occurrence and timing of slow oscillations and the ongoing sensory input that resulted from respiration concluding that there is a strong relation between the timing of respiration.

Needless to say, we can not live without breathing. It is interesting that the olfactory system is closely related to respiration. Most primates require rapid sensing of environmental danger, identification of food or recognition of sex differences; and this sensing is done through the olfactory system enhancing immediate responses to impending events. These behaviors are indispensable for animal life and olfaction is the main sense which produces perception of the external world and gives emotional meaning toward surrounding situations. The human brain is covered by an expanded cerebral cortex that has sensory and motor functions that we are believed to be for higher processing. Fontanini et al. (2004) have suggested that the fundamental neuronal architecture of the cerebral cortex first evolved in the context of the olfactory system and was adapted for uses with the other sensory systems through the evolution of the neocortex. Our results suggest that cortical and thalamic activities in humans are sensitive to respiratory rhythmic input with olfactory information ascending directly to the olfactory and limbic areas; accordingly, breath-by-breath inspiration activates these areas and influences the cortical rhythm.

To live is to breathe continuously. Breathing patterns are affected by listening to the deep structure of the brain and reflecting emotions or feelings. At the same time breathing or smelling affects emotions or feelings.

References

Boiten FA, Frijda NH, Wientjes CJE (1994) Emotions and respiratory patterns: review and critical analysis. Int J Psychophysiol 17: 103-128

Davis M (1992) The role of the amygdala in fear and anxiety. Ann Rev Neurosci. 15: 253-375

Fontanini A, Spano P, Bower JM (2003) Ketamine-xylazine-induced slow (<1.5 Hz) oscillations in the rat piriform (olfactory) cortex are functionally correlated with respiration. J Neurosci 23: 7993-8001

Harper RM, Frysinger RC, Trelease RB, Marks JD (1984) State-dependent alteration of respiratory cycle timing by stimulation of the central nucleus of the amygdala. Brain Res 306: 1-8

Homma S, Musha T, Nakajima Y, Okamoto Y, Blom S, Flink R, Hagbarth KE, Moström U (1994) Location of electric current sources in the human brain estimated by the dipole tracing method of the scalp-skull-brain (SSB) head model. Electroencephalogr Clin Neurophysiol 91: 374-382

Homma I, Masaoka Y, Hirasawa K, Yamane F, Hori T, Okamoto Y (2001) Comparison of source localization of interictal epileptic spike potentials in patients estimated by the dipole tracing method with the focus directly recorded by the depth electrodes. Neurosci Lett 304: 1-4

LeDoux J (1998). *The Emotional Brain:* The Mysterious Underpinnings of Emotional Life. Simon & Schuster, NY.

Mador MJ, Tobin MJ (1991) Effect of alternations in mental activity on the breathing pattern in healthy subjects. Am Rev Respir Dis 144: 481-487

Masaoka Y, Homma I (2001) The effect of anticipatory anxiety on breathing and metabolism in humans. Respir Physiol 128: 171-177

Masaoka Y, Hirasawa K., Yamane F., Hori T, Homma, I (2003) Effects of left amygdala lesion on respiration, skin conductance, heart rate, anxiety and activity of the right amygdala during anticipation of negative stimulus. Behavior Modification 27 (5): 607-619

Masaoka Y, Homma I (2004) Amygdala and emotional breathing. In: Champagnat J (ed) Frontiers in modeling and control of breathing. Kluwer Academic/Plenum Publishers, NY, pp 9-14

Masaoka Y, Nobuyoshi K, Homma I (2005) Inspiratory phase-locked alpha oscillation in human olfaction: source generators estimated by a dipole tracing method (2005) J Physiol 566: 979-997

Morris JS, Friston KJ, Buchel C, Frith CD, Young AW, Calder AJ, Dolan RJ (1998) A neuromodulatory role for the human amygdala in processing emotional facial expressions. Brain 121: 47-57

Steriade M (2000). Corticothalamic resonance, states of vigilance and mentation. Neuroscience 101: 243-276

Reiman EM, Fusselman MJ, Fox PT, Raichle ME (1989) Neuroanatomical correlates of anticipatory anxiety. Science 243: 1071-1074

Ramsay SC, Adams L, Murphy K, Corfield DR, Grootoonk S, Bailey DL, Frackowiak RS & Guz A (1993) Regional cerebral blood flow during volitional expiration in man: a comparison with volitional inspiration. J Physiol 461: 85-101

Breathing Mind in 'Noh'

Ikuo Homma, Yuri Masaoka, and Naohiko Umewaka

Department of Physiology, Showa University School of Medicine, 1-5-8 Hatanodai, Shinagawa-ku, Tokyo 142-8555, Japan

Summary. The expression of Noh, one of the most traditional performing arts in Japan, does not present external expressions or attitudes but interior expressions. We examined how we sympathized their interior expressions without body expression and word, and how the interior expressions are made in the actor's brain. We simultaneously recorded EEG activities and respiratory movements during performing sad emotions in the Noh actors. Their chest walls were hyper inflated and breathing rhythm increased during the performance and negative potential changes were recorded from 300msec before to 300msec after the onset of inspiration. Dipoles, estimated during the pre-inspiratory potential changes using the dipole tracing method, were located in the temporal lobe especially in hippocampus, amygdala and insula. The results suggest that sad emotions in Noh are expressed in increasing of breathing rate and the sources of EEG potential changes related to the respiration are located in limbic and paralimbic areas.

Key words. Amygdala, Dipole tracing method, sadness, EEG, breathing rhythm

1 Introduction

In performing arts, audiences strike a sympathetic chord in actor's attitude or words that imply various emotions such as pleasant, sad and angry. Attitude or external expression is a common transmission for the most performing arts. The expression of Noh, which is one of the traditional performing arts in Japan, does not represent external expressions or attitudes but interior expressions. Noh actors do not express sadness or distress in attitude, even in their face expressions. This type of performing art is very rare in the world and Noh has not changed its form for 600 years (1). How can we sympathize their interior expressions without body expression and words? How the interior expressions are made in the actor's brain and how are they reflected on? Noh experts had been focused on breathing rhythm as an implement of expression of Noh, and believed that mind expression could transmit to breathing rhythm (Umewaka 2001). Can it be possible to inquire the emotional breathing from the brain science?

Several noninvasive methods have been developed to evaluate the source of human brain activities in this decade. fMRI (functional Magnetic Resonance Imaging) represents the local changes of hemodynamics associated with brain activities and MEG (Magnetic Encephalography) represents the magnetic distribution around the scalp produced by neural activities in the brain. Recently EEG Dipole Tracing Method(EEG/DT) has been developed, which evaluates the source localization produced by electric activity generated by synaptic and /or spike activity in the brain (Musha and Okamoto 1999). The location of the equivalent current dipoles is determined by calculations using algorithms minimizing the squared difference between the potentials actually recorded from the scalp and those calculated from the estimated equivalent dipoles. EEG/DT has been available for use in clinics, especially for evaluating the epileptic focus (Homma et al. 2001).

Respiratory movements are generated by the respiratory center in the lower brainstem for metabolic demands. Contrary to the metabolic breathing, breathing can be generated by the respiratory center in the cerebral cortex for behavioral purposes. We showed that respiratory rate increased during the subject felt anxiety and there was a positive correlation between increase of respiratory rate and individual trait anxiety scores (Masaoka and Homma 2001). Recently, Respiratory-related Anxiety Potential (RAP) has been recorded in EEG during anticipatory anxiety in normal subjects. The sources estimated by EEG/DT were located in the temporal pole and amygdale in the limbic system (Masaoka and Homma 2000).

In this work we simultaneously recorded EEG and respiratory movement from 3 Noh players during they were performing a sadness mood in the Noh drama of Sumidagawa. Sumidagawa is one of the typical Noh drama in which the main actor, Shite, performs a woman whose son was kidnapped by a children dealer. She knew hi is her son who died after a long trip with a dealer.

2 Method

2.1 Subjects

Three 'Noh' drama experts were participated. All experts were men without any neurological or respiratory diseases. Subject A was 43 years old and had played 'Noh' for more than 30 years. Subject B was 38 years old and had played for more than 20 years. Subject C was 52 years old and had played also for more than 20 years. They have qualifications for playing 'shitekata' who can play principals.

2.2 Experimental Setup

A subject kept a standing position as the same on the stage of 'Noh' drama 'Sumidagawa' in a shielded room. The EEGs were recorded from 21 surface electrodes attached on the player's scalp according to the international 10/20 systems with the reference electrode on the right earlobe. A grand electrode was placed on the forehead. The EEG was recorded in wireless by EEG telemeter system (WEE 6112, Nihon Koden). EEG was amplified and filtered (band passed: 0.016-200Hz) and stored on the EEG analyzer (DAE-2100, Nihon Koden). Twenty-one electrode positions and reference point positions (nasion, inion, bilateral pre-meatus points and vertex) were measured with a 3 dimensional digitizer (Science 3DL). After the experiment, CT images of the head were obtained from each subject and MRI images were also obtained from subjects with markers putting on the reference points.

Respiratory movements were recorded by respiratory induction plethysmography using a Respitrace transducer (Ambulatory Monitoring, Ardsley, NY). The band of the Respitrace consists of a coil of teflon insulated wire, which was attached around the rib cage just below the axilla. Respiratory rib cage movements changed the inductance of the band, and the inductance changes were converted into proportional voltage changes passed through a band-pass filter (0.007-4Hz). The sensor gain was adjusted so that rib cage movements during breathing gave rise to cyclic voltage variations within a range up to 5V.

Subjects were asked to play a Noh drama 'sumidagawa', performing a sadness of the woman who lost her baby. EEG and respiratory movement were simultaneously recorded during this time and stored all data on the EEG analyzer.

The location of dipoles was estimated by the EEG/DT using with 3-shell (SSB:Scalp, Skull and Brain) realistic head models (BS-navi:Brain Space Navigator, Brain research and Development). SSB realistic head models were reconstructed either from CT images in Subject B and C or MRI images in Subject A. The shapes of the SSB head model are necessary for making a transfer matrix file which taking account for different conductivities of the scalp, skull and brain (0.33, 0.004125, 0.33 s/m respectively). The locations of dipoles were determined when a square difference between potentials calculated from estimated dipoles and potentials actually recorded from the scalp become minimum. The accuracy of the estimation was expressed by means of 'dipolarity', which shows the percentage of the correspondence between the calculated potential and the actual potential. Several studies described methods to approximate intracranial electrical generator sources from EEG recordings. Most methods used least square algorithms to compare actually recorded EEG potentials and theoretically calculated potentials from properly chosen dipoles (Okamoto et al. 1983). Since a poorly conductive skull smears the potential distributions out, a multiple-shell head model is indispensable to take this effect into account. There were several studies for assuming the location of the source using the 3-shell head model (scalp, skull and brain) in comparison with estimating from EEG activities and directly recorded by depth electrodes or subdural electrodes (Homma and Masaoka 2001, Frink et al. 2000). In the previous studies, we used the realistic 3-shell head model in which the conductivity of the scalp, skull and brain was set at 0.33m/s, 0.0041m/sand 0.33m/s, similar to that used in the study of Cuffin et al. (1991).

3 Results

Respiratory rate increased from 15.0 ± 0.0 to 22.0 ± 3.5(mean $\pm$ SD) rate per minute in Subj.A and from 16.0 ± 1.7 to 25.0 ± 1.7(mean $\pm$ SD) in Subj.B during a performance of 'Sumidagawa'. There was no change in the respiratory rate in Subj.C. Typical example of respiratory changes was shown in Fig.1. An increase of chest wall circumference at the end expiratory phase was also obvious in the tracing of the respiratory movement in Fig.1.

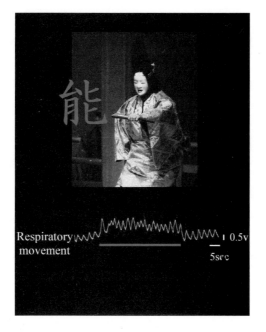

Fig. 1. Performance of Noh with respiratory movement. End expiratory level and respiratory rate increased during performance of 'Sumidagawa'

EEG activities were triggered at the onset of inspiration and added to be averaged. Five EEG activities were averaged during the performance in each subject. EEG potential changes were observed between 450msec before the onset of inspiration to 350msec after the onset of inspiration in Subject A and 450msec before to 200msec after the onset in Subject B (pre-inspiratory potential) (Fig.2). Potential changes were not observed in Subject C. Mean absolute background potential of all channels was 1.1μv in Subject A and 4.9μv in Subject B.

The mean absolute peak potential in the pre-insp potentials was 2.9μv in Subject A and 7.1μv in Subject B (Fig.3). Topography recordings at the peak of the pre-insp potentials were shown in the lower right panel of each subject in Fig.3. One or two equivalent current dipoles were estimated in the right side of the head from the topography recordings in both subjects. The location of equivalent current dipole was estimated with calculations with either 1- or 2-dipole algorithms using BS-navi. The dipolarities were less than 90% in the pre-insp potentials in the 1-dipole algorithm in both subjects. Contrary to the 1-dipole estimation, the dipolarities were more than 97% in the 2-dipole algorithm in both subjects.

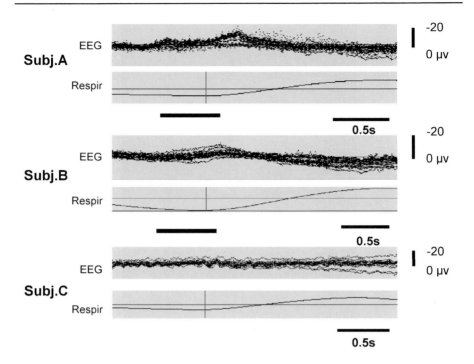

Fig. 2. Averaged EEG and respiratory movements during the performance in 3 subjects. Five onsets of inspiration during the performance were used for averaging EEG. Averaged EEG activities in each electrode were superimposed to show in the upper tracing of each subject. The horizontal bars indicated the potential changes (pre-insp.potential) in Subj. A and B. No potential changes were observed in Subj. C

The locations of dipoles during the pre-insp potentials were shown on 3 dimensional slice images (the axial slice in the left lower panel, the coronal slice in the left upper panel and the saggittal slice in the right upper panel) in Fig.3. Eleven dipoles with dipolarity more than 97% were located in the following areas in Subject A. One dipole was in the right mesial-temporal lobe, five in the right lateral temporal lobe, three in the left mesial temporal lobe and two in the left anterior cingulated cortex. In Subject B, ten dipoles were located in the following areas. One dipole was in the right mesial temporal lobe, two in the right temporal lobe, two in the right lower temporal lobe, one in the right insula, one in the right thalamus, two in the left mesial temporal lobe and one in the occipital lobe. More than half dipoles of them were focused in the right temporal lobe in both subjects.

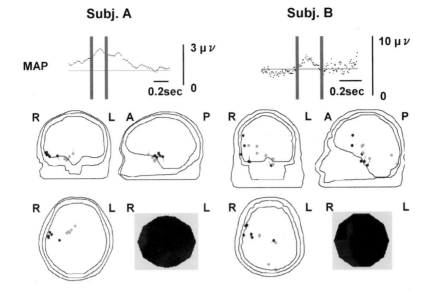

Fig. 3. Mean absolute potentials (MAP) of all channels (*upper*) and dipole locations (*lower*) were shown in subj.A and B. Dipole locations were calculated between two vertical bars in the pre-insp potentials. Two dipoles, pared with red and green dots, were simultaneously estimated and shown in the 3 dimentional slices (coronal, saggital and axial). A topography of the pre-insp potentials was shown in the right bottom of each subject

Dipoles were estimated at every 5msec and those having dipolarities more than 97% were displayed on the CT or MRI slice images. Fig.4 showed the dipoles at different times during the pre-insp potential from 375msec to 275msec before the onset of inspiration in subject A. Upper tracings shows the mean absolute potentials around the pre-insp potential. Dipoles were estimated between 375 to 350msec, which was shown with vertical bars, before the onset of inspiration in the left panel of Fig.4. Dipoles were also estimated between 350 to 325msec in the middle panel and between 325 to 275msec in the right panel of Fig.4. Two dipoles were estimated between 375 to 350 msec before the onset of inspiration. One dipole was focused in the right insula which was shown on the corresponding axial and coronal MRI slices and the other one was focused in the right lateral temporal lobe which was shown on the corresponding coronal MRI slice. Two dipoles were also estimated between 350 to 325msec before the onset of inspiration. One was focused in the left hippocampus, which was shown on the corresponding coronal and axial MRI slices, and the other one was focused in the lower part of the left hippocampus, which was shown on the corresponding coronal MRI slice. Five dipoles were estimated between 325 to

275msec before the onset of inspiration. One dipole was focused in the right bottom of temporal lobe, which was shown on the corresponding coronal and axial MRI slices, and another one was focused in the right amygdale, which was shown on the corresponding coronal and axial MRI slices. Two dipoles were focused in the right lower temporal lobe, one in the right prefrontal cortex, one in the right thalamus and one in the left occipital lobe.

4 Discussion

We focused on two things in the 'Noh' drama actors. One thing was the breathing rhythm. We recorded their breathing pattern in this experiment because of the breathing has been said the expression of their interior emotional changes. The greatest thespians of Noh discovered that the body expression could be altered by breathing and succeeded in incorporating breathing rhythm as an artistic element of their craft (Umewaka 2001).

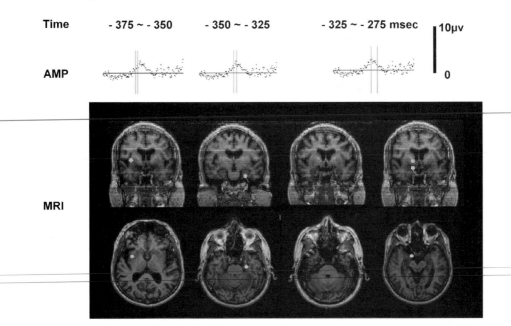

Fig. 4. The mean absolute potential (MAP) and dipole locations, superimposed on the MRI slices, during the pre-insp potential in subj.A . Dipole locations between 375 to 350msec (*left*), between 350 to 325msec (*middle*) and between 325 to 275msec (*right*) before the onset of inspiration

During performing 'Sumidagawa', their breathing rate increased and their lung was hyperinflated. In the human experiment, several works showed that breathing rate increased with the subjects' increasing negative emotions (Masaoka and Homma 1999). Breathing rate increased during the anticipatory anxiety and an increase of breathing rate correlated with individual trait anxiety scores (Masaoka and Homma 2001). The increased breathing rate resulted in hyperventilation and the end tidal CO_2 concentration decreased. This type of breathing is so called behavioral breathing and is distinguished from the breathing for metabolic demand. It is without doubt that the metabolic breathing pattern is generated in the brainstem, especially in the medulla, and the behavioral breathing pattern is generated in the higher center. The other focus was the brain area related to the sad emotion, and the area related increasing respiratory rate associated with sad emotion.

In the EEG recordings, negative potentials were observed from 300msec before to 350msec after the onset of inspiration during the actors' performing a sad affect. There is a recent topic in the animal experiment that pre-inspiratory activity, which is produced by pre-inspiratory neuron, generates a respiratory rhythm (Ballanyi et al. 1999). The pre-inspiratory potential recorded in the present study may be related to the increased breathing rate during the performance. We could estimate several dipoles, which had dipolarity more than 97%, using the 2-dipole model during the pre-insp potentials in two actors. Six out of 11 dipoles and 5 out of 10 dipoles were estimated in the right temporal lobe in the Subject A and Subject B respectively. There were dipoles in the right amygdale and in the lateral and medial part of the right temporal lobe in both cases. There were several studies showing that the amygdale in the limbic system generated negative emotions (LeDoux 2000, Maren 1999). Present data suggests that the source generator of the pre-insp potential may be located in the right limbic and para limbic area. Before the activation of right amygdale, right lateral temporal lobe and left hippocampus were also activated (Fig.4). The data also suggest that amygdalohippocampal circuit may produce the pre-insp potential and play a pivotal role in the negative emotion and emotional breathing.

References

Ballanyi K, Onimaru H, Homma I (1999) Respiratory network function in the isolated brainstem-spinal cord of newborn rats. Prog Neurobiol 59: 583-634

Cuffin BN, Cohen D, Yunokuchi K (1991) Tests of EEG localization accuracy using implanted source in the human brain. Ann Neurol 29:132-138

Frink R, Homma S, Kanamaru A (2000) Source localization of interictaol epilep-
tiform spike potentials estimated with a di;ole tracing method using surface
and subdural EEG recordings. Clin. Neurophysiol 53:1-11

Homma I, Masaoka Y, Hirasawa K, Yamane F, Hori T (2001) Comparison of
source localization of interictal epileptic spike potentials in patients estimated
by the dipole tracing method with the focus directly recorded by the depth
electrodes. Neurosci Lett 304:1-4

LeDoux JE (2000) Emotion circuits in the brain. Annu Rev Neurosci 23:155184

Maren S (1999) Long-term potentiation in the amygdala: a mechanism for emotional learning and memory. Trends Neurosci 22: 561-567

Masaoka Y, Homma I (1999) Expiratory time determined by individual anxiety
levels in humans. J. Appl Physiol 86:1329-1336

Masaoka Y, Homma I (2000) The source generator of respiratory-related anxiety
potential in the human brain. Neurosci Lett 283: 21-24

Masaoka Y, Homma I (2001) The effect of anticipatory anxiety on breathing and
metabolism in humans. Respir Physiol 128:171-177

Musha T, Okamoto Y (1999) Forward and inverse problems of EEG dipole localization. Critical Reviews in Biomedical Eng 27:189-239

Okamoto Y, Teramachi Y, Musha T (1983) Limitation of the inverse problem in
body surface potential mapping. IEEE Trans, 11:749-754

Umewaka N (2001) Noh Theatre, The Aesthetics of Breathing. In:Haruki Y,
Homma I, Umezawa A, Masaoka Y (ed) Respiration and Emotion. Springer,
Tokyo, pp173-175

The Source Generator of Event-Related Potentials with Recognizing Facial Expression by the Dipole Tracing Method

Nahoko Yoshimura[1,3], Yuri Masaoka[2], Ikuo Homma[2], and Mitsuru Kawamura[1]

[1]Department of Neurology and [2]Second Department of Physiology, Showa University School of Medicine, Hatanodai 1-5-8, Shinagawa-ku, Tokyo 142-8555, Japan
[3]Department of Neurology, Tokyo Metropolitan Ebara Hospital, Higashi-yukigaya 4-5-10, Ota-ku, Tokyo 145-0065, Japan

Summary. Studies in human and other primates reveal that the amygdala, orbitofrontal and ventromedial frontal cortices, and cerebral cortex in and near the superior temporal sulcus (STS) region are important components of facial cognition. We investigated neural mechanisms that were involved in recognizing facial expressions in patients with Parkinson's disease (PD) by visual event-related potentials and the dipole-tracing method. Our result showed that dysfunction of the amygdala in patient with PD changes the neural substrates that are normally used to recognize facial expression. Instead of amygdala, STS region in patients with PD were predominantly involved in response to fearful facial expression, as compared with normal subjects.

Key words. Event-related potentials, face recognition, amygdala, dipole-tracing method, Parkinson's disease

1 Introduction

We found previously that patients with Parkinson's disease (PD) were impaired with respect to the recognition of fear and disgust in facial expressions (Kan et al.,2002). Functional neuroimaging has revealed that both the amygdala and striatum are involved in processing expressions of fear (Morris et al., 1996). As there is evidence that the amygdala and striatum

135

do not function normally in patients with PD (Mattila et al., 1999; Ouchi et al., 1999), it is possible that the disturbance of emotional recognition in patients with PD can be attributed to pathological changes in these regions of the brain. We hypothesized that dysfunction of the amygdala in patients with PD changes the neural substrates that are normally used to recognize emotion. To investigate the neural mechanisms that underlie this impairment, we recorded visual event-related potentials (ERPs) in response to the viewing of fearful facial expressions, and determined the location of the equivalent current dipoles (ECDs).

2 Methods

Ten elderly volunteers and 9 patients with PD were studied. Fearful, surprised, and neutral facial expressions were presented randomly for 500 ms with a probability of 0.1, 0.1, and 0.8 at intervals of 1500 ms. Each subject was asked to press a button as soon as a fearful face appeared on the monitor. An electroencephalogram (EEG) was recorded from 19 scalp sites according to the international 10-20 system referred to the right earlobe. The scalp–skull–brain/dipole tracing (SSB/DT) method was applied to the individual EEG topographies to determine the location of components of the ERPs. The SSB/DT method has been used to reliably evaluate neural activity in deep locations, such as the limbic system (Masaoka and Homma, 2000; Masaoka et al., 2003).

3 Results

The ERPs that were elicited in response to the facial stimuli consisted of a negative peak (N1), two positive peaks, and a subsequent slow negative shift. N1 was localized, predominantly, to the bilateral occipitotemporal region. For N1, the equivalent current dipoles were concentrated in the fusiform gyrus, right superior temporal gyrus, parahippocampal gyrus, cingulate cortex, and cerebellum, in normal subjects. In response to the fearful stimulus, dipoles were also generated from the amygdala in seven out of ten normal subjects (Fig.1). In a Chi-square test, this finding was statistically significant. In contrast, in patients with PD, N1 was centered bilaterally in the angular gyrus and supramarginal gyrus, and there was no neuronal activity in the amygdala. After N1, dipoles moved toward the frontal region in normal subjects, whereas they remained in the parietal lobes in patients with PD.

This result suggests that neither the amygdala nor the temporal visual-associated cortices in patients with PD are involved in responding to fearful expressions. Temporal and limbic-related cortices are involved in retrieving emotional information, whereas somatosenrory-related cortices permit the recalling of knowledge about emotion (Adolphs et al., 2003). The parietal somatosensory cortex is preferentially recruited for emotionnal recognition in patients with PD, because both the amygdala and orbitofrontal cortex show relatively less response to facial expressions than is the case in normal subjects. Changes of corticostriatal connection may overcome the mild cognitive emotional deficits that are present in patients with PD.

Fearful face

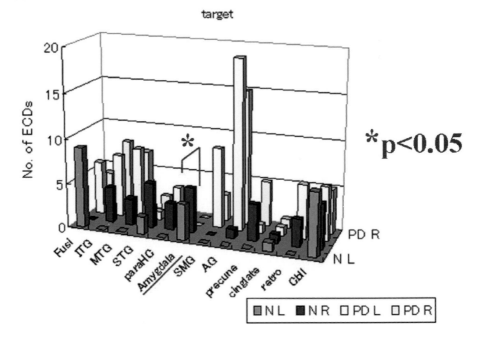

Fig. 1. Localization and number of dipoles estimated for component N1 to fearful face. The Z-axis is the number of ECDs. Two frontal rows indicate left (L) and right (R) hemisphere of normal subjects, and the others, patients with PD

References

Adolphs R, Tranel D, Damasio AR (2003) Dissociable neural systems for recognizing emotions. Brain Cogn 52:61-69

Kan Y, Kawamura M, Hasegawa Y, Mochizuki S, Nakamura K (2002) Recognition of emotion from facial, prosodic and written verbal stimuli in Parkinson's disease. Cortex 38:623-630

Masaoka Y, Homma I (2000) The source generator of respiratory-related anxiety potential in the human brain. Neurosci Lett 283:21-24

Masaoka Y, Hirasawa K, Yamane F, Hori T, Homma I (2003) Effects of left amygdala lesions on respiration, skin conductance, heart rate, anxiety, and activity of the right amygdala during anticipation of negative stimulus. Behav Modif 27:607-619

Mattila PM, Rinne JO, Helenius H, Röyttä M (1999) Neuritic degeneration in the hippocampus and amygdala in Parkinson's disease in relation to Alzheimer pathology. Acta Neuropathol 98:157-164

Morris JS, Frith CD, Perrett DI, Rowland D, Young AW, Calder AJ, Dolan RJ (1996) A differential neural response in the human amygdala to fearful and happy facial expressions. Nature 383:812-815

Ouchi Y, Yoshikawa E, Okada H, Futatsubashi M, Sekine Y, Ito M, Sakamoto M (1999) Alterations in binding site density of dopamine transporter in the striatum, orbitofrontal cortex, and amygdala in early Parkinson's disease: compartment analysis for beta-CFT binding with positron emission tomography. Ann Neurol 45:601-610

Source Localization of Event-Related Potentials Related to Cross-Modal Semantic Interference Effect Using Scalp-Skull-Brain Dipole Tracing Method

Junko Shinoda[1], Kazuyuki Nakagome[1], Masaru Mimura[1], and Ikuo Homma[2]

[1] Department of Psychiatry, Showa University School of Medicine, 1-5-8 Hatanodai, Shinagawa-ku, Tokyo 142-8666, Japan
[2] Second Department of Physiology, Showa University School of Medicine, 1-5-8 Hatanodai, Shinagawa-ku, Tokyo 142-8666, Japan

Summary. The object of this study was to investigate the source localization of event-related potentials (ERPs) related to cross-modal semantic interference effect. Twenty-eight healthy volunteers participated in the study, in which ERPs were measured while subjects judged the semantic relationship between a test picture and a preceding context picture where a superimposed context word was either semantically related (related condition) or unrelated (unrelated condition) to the test picture. The source localization of the equivalent current dipoles (ECDs) of the ERP components, which significantly varied between the related and the unrelated condition when the pictures were semantically unrelated, was determined by the scalp-skull-brain dipole tracing (SSB-DT) method using a two-dipole model. Three distinct components were observed in the difference ERPs obtained by subtracting the ERPs of the unrelated condition from those of the related condition: An early positive component arising in the latency range of 380–420 ms, a negative component in the latency range of 520–560 ms, and a late positive component in the latency range of 680–720 ms. According to the SSB-DT method, significant ECDs were found only for the negative component. The ECDs of the negative component were localized in the right anterior cingulate gyrus and the left superior temporal gyrus, which was nearly in accordance with the source generator of the response conflict negativity reported in previous studies.

Key words. Event-related potential, semantic processing, interference, response conflict, scalp-skull-brain dipole tracing

1 Introduction

Natural environments often confront us with multiple sources of information that could potentially control or guide our behavior. When information from two or more domains conflicts, we often experience interference. The interference effect shares some properties with the inhibitory effect in negative priming. Tipper (1985) demonstrated that the priming effect of an ignored picture on a subsequent categorically related picture was inhibitory and termed this negative priming. Initially, negative priming was interpreted as reflecting active inhibition of internal representations of the ignored object (Neill 1977; Tipper and Driver 1988). More recently negative priming has been related to the response selection stage, which is hampered by confusion between the episodic memory representations of the ignored object and the processing required for the subsequent object (Neill 1997; Stolz and Neely 2001).

Previous studies have demonstrated that certain event-related potentials (ERPs), such as N400, are sensitive to semantic priming effect not only for words (Holcomb, 1988), but also for pictures (Holcomb and McPherson 1994) and cross-modal conditions (words and pictures) (Federmeier and Kutas 2001). If the interference effect occurred at the stimulus identification stage, one could assume that the negative priming effect might affect the amodal semantic processing reflected by N400. Moreover, previous ERP studies using either the Flanker task or the Stroop task have identified ERP components related to response conflict processing, such as N2 (Van Veen and Carter 2002) or N450/sustained potential (West 2003). Van Veen and Carter (2002) demonstrated that N2, occurring prior to the response on correct conflict trials, could be modeled as having a generator in the anterior cingulate cortex. According to the hypothesis of memory confusion during episodic retrieval, the ERP components related to conflict processing may well be associated with the interference effect.

Gernsbacher and Faust (1991) proposed a well-designed task that enabled us to investigate the effect of cross-modal semantic interference between pictures and words. In the task, subjects were required to ignore the word presented in the context display and to verify the semantic association between the pictures presented in the context and test displays. In the present study, in an attempt to verify whether the cross-modal interference effect arises in the stimulus identification or response selection stage, we investigated the ERPs related to the interference effect in the Gernsbacher and Faust task, and also determined the source localization of the equivalent current dipoles (ECDs) using the scalp-skull-brain dipole tracing (SSB-DT) method (Masaoka and Homma 2000).

2 Subjects and Methods

2.1 Subjects

Twenty-eight normal healthy volunteers (16 males, 12 females; mean age, 24.8 ± 5.7 years) participated in the study. Subjects gave written informed consent for participation. All were right-handed, with no history of psychiatric or neurologic illness, and had normal corrected or uncorrected visual acuity. This study was approved by the Ethics Committee of Showa University.

2.2 Procedure

ERPs were recorded while the subjects performed a task proposed by Gernsbacher and Faust (1991). In the task, subjects first viewed a context display (750 ms), which contained a line drawing of a common object and a superimposed word. Following a 500-ms interval, subjects were shown a test display of another (target) picture (750 ms). The subjects' task was to verify whether the target picture was semantically related to the picture shown in the context display. The experiment consisted of 160 trials in total. Half of the trials (80) were assigned to a filler condition in which target pictures were semantically related to the preceding context pictures, whereas target pictures and preceding context words were semantically unrelated. In the other half of the trials (80), the target pictures were semantically unrelated to the preceding context pictures. The target pictures and the preceding context words were semantically unrelated in half of those trials (unrelated condition; 40), while they were semantically related in the other half (related condition; 40). Behavioral interference effect was calculated by subtracting the accuracy and response latencies in the unrelated condition from those in the related condition.

2.3 EEG Recording

ERP recordings began 168 ms prior to the onset of the test display and lasted for 1024 ms. The EEG was recorded from 58 Ag/AgCl electrodes referenced to the left mastoid, and all data were rereferenced off-line to the average of the activity at the left and right mastoids. ERPs were averaged separately for the filler, unrelated, and related conditions, and difference waveforms were obtained by subtracting ERPs for the unrelated condition from those for the related condition.

2.4 Dipole Tracing Analysis

To determine the location of the current source generators of the ERPs, ERP data were analyzed according to the method of Masaoka and Homma (2000) with a Brain Space Navigator (Japan Graphics, Tokyo, Japan), using the standard three-layer head model (scalp-skull-brain) for the grand-averaged difference ERPs, and assuming standard conductivity (0.33 S/m for scalp and brain, 0.0041 S/m for skull). We chose the two-dipole estimation in this study. We evaluated only dipoles within a range of 40 ms of each peak activity, and for which the root-mean square quality of fit (dipolarity) exceeded 98%.

3 Results

3.1 Behavioral Data

Statistical analyses of accuracy and reaction time (RT) revealed that accuracy was significantly lower in the related and filler conditions than in the unrelated condition, and that RT was prolonged significantly in the related condition compared to the other two conditions.

3.2 Event-Related Potential

Visual inspection of the grand-averaged waveforms of the difference ERPs obtained by subtracting those of the unrelated condition from those of the related condition showed a negative component with a latency approximating 540 ms (N540) and a positive component with a latency near 400 ms (P400) and 700 ms (P700; Fig. 1). Statistical analysis revealed a significant main effect of condition (related, unrelated) for P400, N540, or P700.

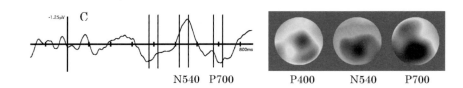

Fig. 1. On the *left* is a representative grand-averaged event-related potential (ERP) difference waveform obtained by subtracting ERPs of the unrelated condition from those of the related condition at Cz. On the *right* are EEG topographies using the mean amplitude data of the difference ERPs calculated within 380–420 ms (P400), 520–560 ms (N540), and 680–720 ms (P700)

3.3 Correlation Between Interference Effect and Event-Related Potentials

We calculated behavioral interference scores by subtracting the accuracy for the related condition from that for the unrelated condition (accuracy interference effect), and the RT for the unrelated condition from that for the related condition (RT interference effect). As for ERP indices, we adopted P400, N540, and P700 amplitudes averaged across all regions. Statistical analyses indicated a significant negative correlation between P400 and an accuracy interference effect (Spearman rho = -0.43, $P < 0.05$), indicating that the larger the P400, the smaller the accuracy interference effect. No significant correlation was obtained between either N540 or P700 and the behavioral interference effect.

3.4 Equivalent Current Dipoles Determined by Scalp-Skull-Brain Dipole Tracing Method

No significant ECDs were found for either P400 or P700, the dipolarity of which exceeded 98%. However, significant ECDs were determined at multiple time points for N540 (520–560 ms). The source generators of N540 were localized in the right anterior cingulate gyrus and the left posterior superior temporal gyrus (Fig. 2).

right   left   right   left   anterior   posterior

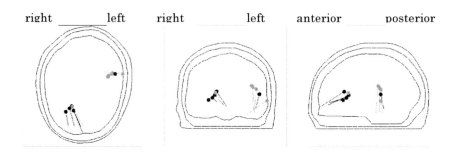

Fig. 2. Locations of dipoles for which the dipolarity exceeded 98% were estimated within the latency range of 520–560 ms

4 Discussion

The present findings carry implications regarding the process underlying the interference effect. Subjects with a smaller interference effect of context words on accuracy, contrary to our assumption, showed a larger amplitude difference (related – unrelated) for P400 than subjects with a larger interference effect. According to our initial hypothesis, the interference effect induced by inhibition of representations of the context word should be associated with a larger amplitude difference between related and unrelated conditions. However, this was not the case. On the other hand, we observed significant negative and positive deflections in the later epochs of the difference ERPs. These negative and positive components appeared at latencies near 540 ms (N540) and 700 ms (P700), respectively, which may well represent the response processing stage subsequent to stimulus identification. Moreover, the SSB-DT method revealed significant ECDs for N540 but not for P700. The source generators were determined in the right anterior cingulate gyrus and the left superior temporal gyrus, which, at least in part, coincides with N2 and N450 (Van Veen and Carter 2002; West 2003). Although the latency of N540 was longer than that of either N2 or N450, the differences in tasks may well explain this discrepancy. In the present study, subjects were required to judge a semantic association, while N2 and N450 were elicited in the Stroop task and the Flanker task, which are less demanding tasks than the one adopted in the present study, since only judgment of the identity of the stimuli was required. Moreover, it can be speculated that the neural activity arising from the left posterior superior temporal gyrus subserves semantic processing involved in the

Gernsbacher and Faust task (Geschwind and Levitsky 1968; Galaburda and Sanides 1980).

Taking into account the ERP components presumably related to conflict processing in the related condition, the interference effect of cross-modal semantic priming of pictures and words might have occurred in the response selection process, presumably because of the conflict in response selection. These findings contradict the initial view, which advocates the relevance of active inhibition of the distractor stimuli to negative priming, suggesting instead that negative priming may arise from confusion between memory representations of the prime trial and the processing required in the probe trial. In the present study, a conflict between the semantic relationship of the context picture/word and the target picture in the related condition might have led to decreased accuracy compared to performance in the unrelated condition.

The present study highlights the advantage of using ERPs with the SSB-DT method for obtaining information complementary to that obtained from behavioral and neuropsychological approaches in elucidating the nature of interference, a phenomenon essential for processing the information from multiple sources we meet in our daily activities.

References

Federmeier KD, Kutas M (2001) Meaning and modality: influences of context, semantic memory organization, and perceptual predictability on picture processing. J Exp Psychol Learn Mem Cogn 27:202-224

Galaburda AM, Sanides F (1980) Cytoarchitectonic organization of the human auditory cortex. J Comp Neurol 190:597-610

Gernsbacher MA, Faust ME. (1991) The mechanism of suppression: a component of general comprehension. J Exp Psychol Learn Mem Cogn 17:245-262

Geschwind N, Levitsky W (1968) Human brain: left-right asymmetries in temporal speech region. Science 161:186-187

Holcomb PJ (1988) Automatic and attentional processing: an event-related brain potential analysis of semantic priming. Brain Lang 35:66-85

Holcomb PJ, McPherson WB (1994) Event-related brain potentials reflect semantic priming effect in an object decision task. Brain Cogn 24:259-276

Masaoka Y, Homma I (2000) The source generator of respiratory-related anxiety potential in the human brain. Neurosci Lett 283:21-24

Neill WT (1977) Inhibitory and facilitatory processes in attention. J Exp Psychol Hum Percept Perform 3:444-450

Neill WT (1997) Episodic retrieval in negative priming and repetition priming. J Exp Psychol Learn Mem Cogn 23:1291-1305

Stolz JA, Neely JH (2001) Taking a bright view of negative priming in the light of dim stimuli: further evidence for memory confusion during episodic retrieval. Can J Exp Psychol 55:219-230

Tipper SP (1985) The negative priming effect: Inhibitory effects of ignored primes. Q J Exp Psychol A 37:571-590

Tipper SP, Driver J (1988) Negative priming between pictures and words in a selective attention task: evidence for semantic processing of ignored stimuli. Mem Cognit 16:64-70

Van Veen V, Carter CS (2002) The timing of action-monitoring processes in the anterior cingulate cortex. J Cogn Neurosci 14:593-602

West R (2003) Neural correlates of cognitive control and conflict detection in the Stroop and digit-location tasks. Neuropsychologia 41:1122-1135

Key Word Index

DISCARDED

CONCORDIA UNIVERSITY LIBRARIES
CONCORDIA UNIV. LIBRARY
GEORGES P. VANIER LIBRARY LOYOLA CAMPUS